POMEGRANATE

POMEGRANATE

Production, Postharvest, Marketing and Export

S.K. Tyagi
Scientist (Horticulture)
Krishi Vigyan Kendra, Khargone
Madhya Pradesh

Akhilesh Tiwari
Senior Scientist (Horticulture)
Research & Training Centre (JNKVV)
Garhakota, Sagar, Madhya Pradesh

NEW INDIA PUBLISHING AGENCY
Pitam Pura, New Delhi – 110 034

NEW INDIA PUBLISHING AGENCY
101, Vikas Surya Plaza, CU Block, LSC Market
Pitam Pura, New Delhi 110 034, India
Phone: + 91 (11)27 34 17 17 Fax: + 91(11) 27 34 16 16
Email: info@nipabooks.com
Web: www.nipabooks.com

Feedback at feedbacks@nipabooks.com

ISBN : 978-93-86546-13-5

Composed, Designed and Printed in India

Preface

Pomegranate (*Punica granatum* L.) is one of the ancient fruits associated with several human cultures of the world. Its nutritional, therapeutic and ornamental values were known to humans since antiquity. Although pomegranate is an ancient fruit plant, it has gained commercial significance recently. Now-a-days there is increasing worldwide demand for this fruit owing to its superior pharmacological and therapeutic properties. Since the pomegranate tree is highly adaptive to a wide range of climate and soil conditions, it is grown in different tropical and subtropical regions.

India is one of the largest producers of pomegranate in the world. During 2013-2014, pomegranate was cultivated over 1.31 lakh ha with an annual production of 13.46 lakh tonnes and productivity of 10.27 tonnes/ ha. At present, Maharashtra is the leading state in acreage covering about 68.80 per cent of the area under pomegranate. Similarly around 70.22 per cent of total production comes from Maharashtra. The other important states are Karnataka, Gujarat and Andhra Pradesh, Telangana and Madhya Pradesh. India is the only country in the world where pomegranate is available throughout the year.

In the present text, proven technologies and procedure for export have largely been compiled for the end-users. The book has six broad chapters, with major focus on production technology, integrated disease and insect pest management (IDIPM), post harvest management and procedure for export of pomegranate.

For the manuscript the information has been gathered from print as well as electronic media with due acknowledgement. I hope that the information contained in this publication will benefit the pomegranate growers, researchers, students, extension workers, processors, exporters, various developmental agencies and other stake holders involved in the promotion of the pomegranate industry in India. The book has been compiled on the basis of available information for guidance and not for legal purposes. Readers are also advised to consult the latest recommendation for pesticides/herbicides before its use. However, I have put my best efforts in preparing this manuscript, but if any error or whatsoever has been skipped out, please accept our apologies and will by heart welcome your suggestions.

S.K. Tyagi
Akhilesh Tiwari

Contents

1

Pomegranate

1.1 Introduction

Pomegranate (*Punica granatum* L.) is a well-known table fruit of tropical and subtropical regions of the world. The Romans received it from Carthage, hence the name of the genus *Punica.* Some botanists place it in the family Lythraceae, of the peculiar type of fruit, called as balausta, most authorities make it the only genus in the family Punicaceae. It belongs to genera *Punica* and family Punicaceae (Chatterjee & Randhawa, 1952; Joshi, 1956). It is a small tree, measuring less than 4 m when cultivated, although it can reach 7 m in the wild. Recent research findings corroborate traditional use of pomegranate as a medical remedy as all parts of this plant have several bioactive metabolites (Holland *et al.* 2009). Consequently, its demand has increased tremendously not only in the western world but also in other parts of the globe. It is claimed to have originated in Iran and perhaps surrounding areas from where it spread to different regions. It was cultivated in ancient Egypt and in early Greece and Italy. The fruit was very popular in Iraq. Over a time, it spread into Asia (Turkmenistan, Afghanistan, India, China, etc.), North Africa and Mediterranean Europe. Interestingly, its domestication process took place independently in various regions (Evernioff 1949; Zukovskij 1950). It is estimated that pomegranate was introduced into India and China over the Silk Road and ultimately into Japan (Bretschneider 1935). However, it was introduced in the Indian Peninsula from Iran during the I[st] century AD and was found growing in Indonesia in 1416 AD. The Spanish sailors brought pomegranates to the New World and some missionaries introduced pomegranate in Mexico and California (Goor and Liberman 1956; Morton 1987). The ability of pomegranate plants to adjust to variable climatic conditions is reflected in the wide distribution of the wild form throughout Eurasia to the Himalayas. The fruit which was described by the Greek botanist Theophrastus about 2350 years before the present (BP) and is mentioned in many Greek and Turkish myths. It is one of the first five domesticated edible fruit crops along with fig, date palm, grape and olive. It has been the symbol of health, fertility and rebirth as mentioned in many ancient cultures. This fruit crop is known to have been domesticated in the Middle East about 5000 years ago (Adsule and Patil 2005). The present scientific name *Punica granatum* is derived from the name *Pomum* (apple) *granatus* (grainy), or seeded apple. In ancient Egypt, pomegranate received the name "*Arhumani*". The Romans first called this species "*malum punicum*" (punic apple or apple of Carthage) that evolved to "*Punicum granatum*". C. Von Linne finally gave the name *Punica granatum* (Hodgson 1917; Bretschneider 1935; Hunt 1989). It is considered an excellent fruit crop for growing in arid zones for its tolerance to drought conditions. Now, it is widely cultivated in Mediterranean, tropical and subtropical regions. However, it is cultivated in Central Asia and to some extent in the USA (California), Russia, China and Japan for fruit production and is also developed as an ornamental tree in East Asia (Mars 1996; Tous and Ferguson 1996).

Pomegranate has several salient features unique to its credit. Pomegranate has ability to withstand harsh and hostile climate. It has built-in capacity to withstand heat, drought and moisture deficit. The versatile adaptability, hardy nature, low maintenance cost, steady but high yields, better keeping quality, fine table and therapeutic values and possibilities to throw the plant into rest period when irrigation potential is generally low, indicate the avenues for increasing the area under pomegranate in India. It has immense medicinal and nutritional value. Pomegranate is one of the richest source of antioxidants.

This ancient fruit has emerged as a commercially important fruit in the recent times. There is a growing demand for good quality fruits both for fresh use and processing into juice, syrup, squash, wine, besides *anardana*, an acidulant product (Pruthi and Saxena 1984). The versatile adaptability, hardy nature, low maintenance cost, steady but high yields, better keeping quality, fine table and therapeutic values and possibilities to throw the plant into rest period when irrigation potential is generally low, indicate the avenues for increasing the area under pomegranate in India.

1.2 Indian Scenario

India is one of the largest producers of pomegranate in the world. During 2013-14, pomegranate was cultivated 130.8 thousand ha with an annual production of 1345.7 thousand tonnes and productivity of 10.3 tonnes/ ha in India (Table 1.1). At present, Maharashtra is the leading state in acreage covering about 68.7 per cent of the area under pomegranate. Similarly around 70.2 per cent of total production comes from Maharashtra. The other important states next to Maharashtra with respect to pomegranate cultivation are Karnataka, Gujarat, Andhra Pradesh, Telangana and Madhya Pradesh.

Table 1.1 : Area, production and productivity of pomegranate in India (2013-14)

State	Area (000'ha)	Production (000'MT)	Productivity MT/ha
Maharashatra	90.0	945.0	10.5
Karnataka	16.6	134.2	8.1
Gujarat	9.4	99.3	10.6
Andhra Pradesh	6.0	90.0	15.0
Telangana	1.7	26.0	15.0
Madhya Pradesh	2.4	25.3	10.6
Tamil Nadu	0.4	13.1	32.7
Rajasthan	0.9	5.6	6.2
Others	3.4	7.2	2.2
Total	130.8	1345.7	10.3

Source : Indian Horticulture Database – 2014

Important pomegranate growing areas in India

The important pomegranate growing states (Figure 1.1) in India are listed below:

Maharashtra: Mainly in Solapur, Nashik, Sangli, Satara, Ahmednagar and Pune and to a limited extent in Osmanabad, Jalna, Beed, Aurangabad, Jalgaon and Dhule districts.

Karnataka: Bijapur, Bagalkot, Koppal, Belgaum, Gadag, Bellary, Raichur Tumkur, Chitradurg and Davanagere districts.

Andhra Pradesh: Anantpur.

Telangana: Mahabubnagar.

Gujarat: Kutch, Banaskantha, Ahmedabad, Sabarkantha and Bhavnagar.

Rajasthan: Hanumangarh and Ganganagar.

Tamil Nadu: Salem, Coimbatore and Periyakulam.

Himachal Pradesh: Solan, Kullu, Sirmour.

Fig. 1.1 : Pomegrante growing states of India

1.3 Global Status

Although, the exact figures of area and production in the world are not available but as per approximate global estimates (Table 1.2), pomegranate production is around 20.55 lakh tonnes from an area of 3.00 lakh ha. At the global level, India is the world's largest producer of pomegranates (743.1 thousand tonnes) followed by Iran (650 thousand tonnes), Turkey (218 thousand tonnes), USA (100 thousand tonnes) and; Afghanistan and Spain (60 thousand tonnes each) during 2011-12. During 2013-14, it was produced over 1.31 lakh ha with an annual production of 13.46 lakh tonnes and productivity of 10.27 t/ha in India.

Table 1.2 : Global area, production, productivity and export of pomegranate (2011-12)

Country	Area (1000 ha)	Production (1000 MT)	Productivity (MT/ha)	Export (x 1000 MT)	Per cent Export share	
					Country wise	Global Market
India*	107.3	743.1	6.9	30.1	4.05	22.75
Iran	60.0	650.0	10.8	60.0	9.23	45.35
Turkey#	8.0	218.0	27.25	86.14	39.51	39.84
Afghanistan**	8.0	60.0	7.5	3.0	5.0	2.26
USA	6.0	100.0	16.7	17.0	17.0	12.84
Israel	1.6	20.0	12.5	4.0	20.0	3.02
South Africa	1.3	4.0	3.1	-	-	-
Spain	3.0	60.0	20.0	14.0	23.33	10.58
Tunisia	2.6	25.0	9.61	2.0	8.0	1.51
Australia	0.25	1.0	4.0	-	-	-
Argentina	0.80	2.0	2.5	-	-	-
Others	101.15	616.9	6.09	-	-	-
Total	300.00	2500.00	9.36	216.2	-	-

Source: International Trade Probe, Jan.31, 2011. Published by the National Agricultural Marketing Council (NAMC) in cooperation with the department of Agriculture, Forestry & Fisheries, Republic of South Africa. *Chronica Horticulture*, 48 (3). 2008, 12-15.

* Indian Horticulture Database (2010-11)

** International Symposium on Pomegranate and other Minor Mediterranean Fruits, Dharwad, India, 2009.

Republic of Turkey –Ministry of Economy, 2012. Fresh Fruits and Vegetables (production and export).

Pomegranate productivity of different countries

Although India is largest producer of pomegranate in the world (Table 1.2), its productivity (6.9 t/ha) is far below than Turkey (27.25 t/ ha), Spain (20.00 t/ha), USA (16.7 t/ha) and Israel (12.5 t/ha) during 2011-12. However, the productivity was increased during 2013-14 (10.27 t/ha).

1.4 Major Pomegranate Exporting Countries

The major pomegranate exporting countries are Turkey (86,100 t), Iran (60,000t), USA (17,000t), Spain (14,000t) and Israel (4,000t) (Table 1.2). Despite a surge in the volume of exports from India over the years, the country exported only 4.05 per cent of its total produce (2011-12) which is quite below the export figures of other countries like Turkey (39.5%), Spain (23.2%), Israel (20.0%) USA (17.0%), and Iran (9.23%). During 2012-13, India has exported 35,000 tonnes of pomegranate. However, India occupies third position in terms of global export share (22.75%) after Turkey (39.84%) and Iran (45.35%). Exports from India are mainly to ASEAN, Gulf, European Union and Pacific Rim countries (Japan, South Korea, China), USA and Canada. During the year 2011-12, India exported 30,150 tonnes of pomegranate valued at Rs 1472.68 million. India has witnessed growth of 532.07 per cent in its pomegranate export over a period of 10 years (2001 to 2011) but it is only 4.05 per cent of its total production which is quite below the figures of other countries.

2

Production Technology

2.1 Soil and Climate

Pomegranate grows well in sandy-loam and well-drained soils having a pH of 7.5. It can also be grown on light soils. Quality and colour development in light soils is good but poor in heavy soils. It tolerates salinity up to 9.00 EC/mm and sodicity 6.78 ESP. Even it can be grown well in slightly saline soils as it is considered a saline-tolerant plant (Patil and Waghmare 1983; Rao and Khandelwal 2001; Ram Asrey *et al.*, 2002). Accumulation of salts, in excess of 0.5% of the soil mass causes dying off of growing roots. Interestingly, application of Paclobutrazol (250 ppm) was reported to reduce salinity damage (Saeed 2005). The presence of water-soluble salts like sulphates and chlorides and exchangeable sodium negatively affect root formation. It can tolerate soil salinity due to the ability of its root system which can accumulate the majority of toxic salts present in the soil and thus prevent their intensive flow-out to the above-ground organs (Marathe *et al.,* 2009). It is well documented that pomegranate tissues accumulated sodium, chlorine and potassium in response to irrigation with saline water (Doring and Ludders 1987; Naeini *et al.,* 2004, 2006). 'Perhaps under such situation, xylopodium, a wood like underground stem performing a function of vegetative reproduction and accumulation organ, acts as a buffer. Intensive absorption of toxic sodium ion, reduces the content of potassium and phosphorus, protein and phospho organic compounds, nitrogen supply from the soil and utilization of phosphorus compounds as a result the content of chlorophyll carotenoids, sugars and tanning substance are reduced in the plant. This leads to decline in crop yield and fruit quality'

Basically, pomegranate is a light-loving plant and reacts negatively to excessive shading (Chadha 2005). At the same time, direct sun-light and considerable heating often causes harmful effect on fruits leading to sun-burns (Sharma *et al.* 2006). However, best quality fruit are produced in arid regions having a long, hot and dry summer. It can easily with- stand temperatures up to 45-48°C in combination with dry hot winds. It is well known that pomegranate is not frost resistant and thus cannot tolerate temperatures lower than 18°C. For its higher fruit yield, regular irrigation is required. Even it hardly tolerates excessive water (Badizadegan, 1975; Kulenkamp *et al.,* 1985; Levin, 2006a) and high soil moisture may lead to wilt disease in India (Sharma *et al.,* 2006).

2.2 Improved Varieties

Alandi

The variety is named after the name of the village it was grown extensively. The fruits are medium in size and the seeds are blood red in colour. The seeds are hard and as such, less preferred by the consumers. It gives fewer yields as compared to other varieties.

GBG-1

This variety is developed by selection method from Alandi. It was released for commercial cultivation in 1936 and renamed as **Ganesh** in 1970. It is a prolific bearer, fruit medium- sized, with yellow, smooth surface and red ting. The seeds are soft with pinkish. It is the commercial cultivar of Maharashtra. The average yield ranges from 8-10 kg per tree.

Mridula

A seedling selection from an open-pollinated progeny raised from the F_1 progeny of a cross Ganesh X Gul- e Shah Red. Its fruits are medium- sized, rind smooth, dark red in colour. It has blood-red arils with very soft seeds, juicy and sweet in taste.

Jyoti

This is a selection from mixed population of Bassein Seedless and Dholka. The fruits are medium to large-sized with attractive colour having dark red arils. The seeds are very soft with high pulp and juice contents. Fruits are borne on the inner side of the canopy and thus do not get damaged due to sun scorching.

Ruby

It is a multiple hybrid of pomegranate developed for aril colour and seed mellowness. A fully grown plant of Ruby is about 2.72 m tall and less spreading (4.7 m2). The mature fruits resemble in size and shape the fruits of Cv. Ganesh, however, the skin colour of Ruby is reddish brown with green streaks. The rind is thin (0.24 cm), fruit contains red bold arils (37.2 g/100 arils) which yield

(72%) thick (1.83 cps) dark red (4.71 at 540 nm) sweet (16°B) juice having tannin content of 173 mg per 100 ml juice. Under Bangalore conditions its performance is comparable with Cv. Ganesh in respect of seed softness (2.19 kg/cm2), fruit weight (270 g) and yield (16-18 tonnes/ha). Better adoptability to milder climate.

Fig. 2.1 : Arakta

Fig. 2.2 : Bhagwa

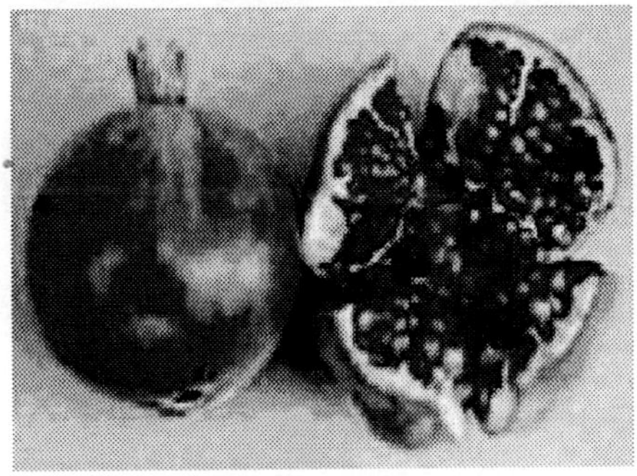

Fig. 2.3 : Mridula

Fig. 2.4 : Ruby

Fig. 2.5 : Dholka

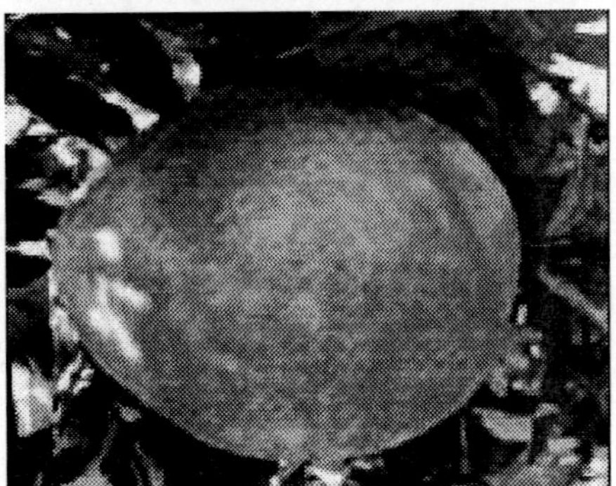

Fig. 2.6 : G-137

Improved varieties of pomegranate (See colour version on page 171)

Musket

This variety is largely grown in Kohar Rahuri region of Ahmednagar district in Maharashtra from seedlings and as a result there is a lot of variation. The seeds are light pink in colour. The fruits are tasty. It is a good cropper.

Dholka

The fruits are big in size. The flesh is pinkish white, seeds are soft, but the juice is acidic. It is a medium cropper. In South India, four varieties such as Paper Shell, Spanish Ruby, Musket and Vellodu have shown considerable promise at Kodur (Andhra Pradesh).

Bhagwa

This variety of pomegranate is heavy yielder and possesses desirable fruit characters. The fruits are matures in 180-190 days after blooming (anthesis) with an average yield of 30.38 kg fruits/tree. Bigger fruit size, sweet, bold and attractive arils, glossy, very attractive saffron coloured thick skin makes it suitable for distant markets.

Fig. 2.7 : An attractive fruit of Bhagwa variety
(See colour version on page 172)

Fig. 2.8 : Arils colour of different varieties of pomegranate
(See colour version on page 172)

Phule Arakta

The variety is a medium yielder and possesses desirable fruit characters. The fruits are bigger in size, sweet with soft seeds, bold red arils. It also possesses glossy, attractive, dark red skin. It is less susceptible to fruit spots and thrips. Hence, the 'Phule Arakta' variety is released for the cultivation in pomegranate growing areas of Maharashtra. It gives high yield (25-30 kg/tree) in case of better management.

2.3 Propagation

Air-Layering

Air layering is carried out in the rainy season. In this method a 1-2 year old, healthy, vigorous, mature shoot of 45-60 cm in length and pencil thickness is selected. A circular strip of bark about 3 cm wide just below a bud is completely removed from the selected shoot. Rooting hormones like IBA (2000-3000 ppm) in Lanoline paste are applied over this portion. Moist sphagnum moss is packed around this portion and tied with polyethylene sheet to prevent the loss of moisture. Application of such hormones promotes early rooting. Light brown roots are visible through the polythene wrap in the month of August- September. The rooted shoot is slowly detached by giving 2-3 successive cuts over a period of week before finally detaching from the parent plant. The polythene sheet is removed before planting them in pots. They are planted in pots and kept in nursery under shade. Top of the shoot is cut back to maintain a proper ratio of leaves: roots. The grafts can be transplanted in the field in the month of September/October.

Fig. 2.9 : Air- layering
(See colour version on page 172)

Plant tissue culture

Plant tissue culture is a technique by which single cell or tissue can

Fig. 2.10 : Plant tissue culture unit

be regenerated in to a complete plant. This process is carried out in a suitable culture medium under controlled environmental conditions of temperature, light and humidity. The advantage of PTC is that a plant can be developed from any part of the mother plant. However, meristematic tissue such as shoot tip is considered to be ideal. Tissue culture means micro-propagation of tissues of the selected Elite plants. The process consists of various steps such as Initiation, Multiplication, Shooting & rooting, Initial acclimatization is done in green houses and Secondary Hardening in shade houses where there is limited control of environmental conditions.

Quality planting material plays an important role in production, productivity and quality in all crops including pomegranate. There are number of disease and pest reported on pomegranate that reduces productivity from 10-90%. In general, vegetative propagated plants have more risk in carrying disease and pest along with the planting material if it is propagated conventionally than seed propagated saplings since the mother plant is exposed to the open environment for many years where there is a high risk of infecting the source plants. Conventional methods of propagation have their own limitations to produce clean and disease free planting material on mass scale. Tissue culture is a hi-tech propagation systems that allows producing clean, disease free, healthy, vigorous and genetically pure planting material on large scale within a short time. The technology offers an advantage of producing large planting material from few mother plants.

Fig. 2.11 : Single node segment proliferated into multinodal shoot
(See colour version on page 173)

2.4 Planting

Season and Spacing

Planting should be done at the beginning of monsoon or by the end of monsoon. Planting distance should be maintained depending upon soil type and climate. A spacing of 4-5 m on marginal and very light soil is recommended. The high-density plantation 5mX2m (1,000 plants/ha), 4.5mX3m (740 plants/ha), 5mX4m (500 plants/ha) gives 2-2.5 times more yield than the normal planting distance (5mX5m) in Deccan plateau.

Pit digging

Pits of 60 cm3 are dug as per the layout plan and exposed to sunlight for one month. Just before filling pits drench the bottom and sides of the pit with 4-5 liters of 0.4% Chlorpyriphos solution. Dust the pits with bleaching powder (a.i. 33% Cl) @ 100g/pit before filling. The pits are filled with top soil mixed with 10 kg of FYM, 1 kg of vermicompost, 1 kg of single super phosphate, 0.5 Kg of neem cake, *Trichoderma* formulation 25g, phosphate solubilising bacteria 25g, *Pseudomonas fluorescens* 25g, *Azotobacter* formulation 25g and *Azospirillum* formulation 25g and watered immediately.

Planting

At the time of planting a small pit sufficient to accommodate the soil ball should be excavated in the centre of the pit. The polythene bag should be removed without disturbing the soil ball. The graft along with the soil ball is carefully lowered in the pit. Soil is lightly pressed around the main stem to remove the air gaps. Plants are watered and adequately supported by stakes.

Fig. 2.12 : Ideal plant geometry
(See colour version on page 173)

2.5 Nutrient Management

Pomegranate is a hardy fruit plant, growing successfully in low fertile soils. Its productivity is greatly increased by the application of manures and fertilizers. Both macro and micronutrients affects its growth, development and productivity.

- About a week after defoliation when all or 85-90% leaves fall down, apply FYM and NPK as detailed below, depending on the age of plant. Remember to adjust the total NPK dose based on the soil and leaf tissue analysis values.
- Apply N and K_2O in 3 split doses, starting at the time of first irrigation after bahar treatment and next at 3-4 weeks interval. Full dose of P_2O_5 should be applied as single dose with first irrigation.
- The manures and fertilizers should be applied in a shallow circular trenches, 30-40 cm away from main stems below tree canopy not beyond 8-10cm depth and cover with top soil and irrigate immediately.
- Nitrogen needs to be applied preferably through urea in black soils and calcium ammonium nitrate (CAN) in red soil, phosphorus as single super

phosphate (SSP) and potash through muriate of potash (MOP). At least 1/3rd fertilizers required should be applied as organics, 1/3rd through inorganics and remaining 1/3rd through fertigation. NPK application should be based on soil test and leaf tissue analysis values.

- Micronutrients (Zn, Fe, Mn and B each 25g per plant or based on plant or soil analysis) should be supplied along with FYM.
- Apply vermicompost @ 1kg/plant, neem cake@500g/plant and phorate @ 20g/plant for controlling nematodes.

Table 2.1 : Quantity of different fertilizers to be applied depending on age and source

Age of plants (year)	FYM (Kg)	Nitrogen (g)			Phosphorus (g)			Potassium (g)		
		N Req.	Source Urea 46% N	CAN 25% N	P Req.	Source SSP 16% P	DAP 46% P	K Req.	Source MOP #18% N	60% K
1.	10	250	540	1000	125	780	1700	271	125	210
2.	20	250	540	1000	125	780	1700	271	125	210
3.	30	500	1090	2000	125	780	1700	271	125	210
4.	40	500	1090	2000	125	780	1700	271	250	420
5. and above	50	625	1360	2500	250	1560	3400	544	250	420

Quantity to be subtracted from Urea/CAN dose if DAP is source of phosphorous.

- When flowering starts schedule 15 drip application of fertilizer N:P:K:: 12:61:00@ 8kg/ha/application on alternate days. This provide 19.46g of N and 98.92g of P_2O_5 per plant
- When fruit setting initiates 15 drip application of fertilizer N:P:K:: 19:19:19: @ 8kg/ha/application, on alternate days. This provides 30.80g of each of the nutrients N, P_2O_5 and K_2O per plant.
- At 100 percent fruit set starts, again schedule 15 drip applications of fertilizer N;P;K:: 0:52:34 or mono-potassium phosphate @2.5 kg/ha/ application, on alternate days for a month. This provides 26.35g of P_2O_5 and 17.23g of K_2O per plant.
- One month before harvesting, schedule 2 drip applications of calcium nitrate@12.5 kg/ha/application at 15 days interval.

2.6 Water Management

Regular irrigation is required during establishment of the plant. Once the plant is established it requires weekly irrigation in summer and bi-weekly in winter. Irrigation is also essential between flowering and fruit ripening, as moisture stress leads to flower and fruit drop and fruit cracking at mature stage. Drip irrigation can also be followed to economize water.

Fig. 2.13 : Drip irrigation for better yield

(See colour version on page 173)

- Drip irrigation with four drippers placed in four directions need to be employed.
- Irrigate the crop immediately after fertilizer application in the case of soil application with light irrigation initially and then irrigate at regular intervals.
- Irrigate the plants depending on water requirement of the plant in different seasons as detailed below.

Table 2.2 : Water requirement of pomegranate plant

Cropping season	Month	Water requirement (Litres/day/plant) will vary with stage of plant
Ambe	January	17
	February	18
	March	31
	April	40
	May	44
Mrig	June	30
	July	22
	August	20
Hasta	September	20
	October	19
	November	17
	December	16

2.7 Weed Management

Pomegranates are grown primarily in arid regions and require regular irrigation. Irrigation also encourages weed growth that competes with trees for water and soil nutrients. In addition, weeds can host a wide array of damaging pomegranate pests and might disturb efficient pest control. Weed control in pomegranates is particularly important in younger trees. Later on, as the trees mature, the shade will inhibit the weeds, and the intensity of weed management procedures is reduced. Weed control should be modified according to weed composition in the specific orchard.

Several herbicides are used to control weeds in pomegranate orchards. Oxyfluorfen is used as pre-emergence weed killers. Both are used to control a wide spectrum of annual broadleaf weeds and grasses. Pos-temergance, Roundup (Glyphosate) is used as a non-selective herbicide to control grasses and weeds while 2,4-dichlorophenoxyacetic acid (2,4-D) derivatives (phenoxy herbicides) are used to control broad leaf weeds. The combination of both post-emergent materials is used when needed (Blumenfeld et al. 2000). Extensive use of these herbicides may result in tree damage; therefore, care should be taken when using them. Bucsbaum et al. (1982) reported phytotoxicity when herbicides were applied to pomegranate grown in pots but not when applied to five year old pomegranates grown in the orchard. These authors found that the best weed control was obtained with Simazine, oryzalin (sulfonamide herbicide, a selective preemergence herbicide used for control of annual grasses and broadleaf weeds) and terbutryne (s-triazine herbicide, a selective herbicide used for control of annual grasses and broadleaf weeds) (Bucsbaum et al. 1982). Usage of Glyphosate to control *Cuscuta monogyna* L., which heavily infested pomegranate trees in Iran, was reported by Saied et al. (2003). Lack of experience growing pomegranate in many areas and small area in cultivation limit both registration and recommendations for herbicide applications to pomegranate orchards. The availability of efficient herbicides is currently relatively limited, and work should be done to broaden the spectrum of certified chemicals for use in pomegranate orchards.

1. Bermuda grass
***Coynodon dactylon* (L.) Pers. (Poaceae)**

2. Purple nut sedge
***Cyperus rotundus* L. (Cyperaceae)**

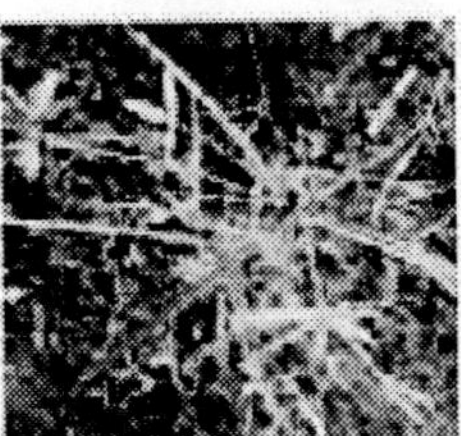

3. Flat sedge
Cyperus iria
(Cyperaceae)

4. Tropical spiderwort
Commelina benghalensis L.
(Commelinaceae)

5. Swine cress
***Didymus* (L.) Sm.**
(Brassicaceae)

6. Annual brachiaria
***Brachiaria deflexa* (Schumach.)**
Robyns (Poaceae)

7. Horse purslane
Trianthema portulacastrum
L. (Aizoaceae)

8. Viper grass
***Dinebra retroflexa* (Vahl.) Panzer.**
(Poaceae)

9. Field bindweed
***Convolvulus arvensis* L.**
(Convolvulaceae)

10. Black nightshade
***Solanum nigrum* L.**
(Solanaceae)

11. Common cocklebur: *Xanthium strumarium* L.
(Asteraceae)

12. False amaranth
***Digera arvensis* Forsk.**
(Amaranthaceae)

13. Puncture vine
***Tribulus terrestris* L.**
(Zygophyllaceae)

14. Asthma herb
***Euphorbia hirta* L.**
(Euphorbiaceae)

15. Carrot grass
Parthenium hysterophorus
L. (Asteraceae)

Fig. 2.14: Common weeds in pomegranate orchard (See colour version on page 174)

A common practice today in modern pomegranate orchards in India is to use polythene mulches. Such mulches conserve soil moisture, reducing water consumption by 20% to 25% (Aulakh and Sur 1999; Ravid *et al.* 2004), and significantly reduce weed population by 20% to 26% compared to controls (Aulakh and Sur 1999; Ravid *et al.* 2004).

Fig. 2.15 : Plastic mulch for water and weed management (See colour version on page 175)

Intercropping

Growing of intercrops in pomegranate orchard is highly profitable as the tree takes about 2 to 3 years to come to good bearing. Vegetable and fodder crops can be successfully grown. Suitable vegetables are cabbage, cauliflower, beans, pea, tomato, carrot, onion, radish, potato and brinjal. Fodder crops such as berseem, lucerne and legume crops like cowpea, moong, urd etc can also be grown. The growing of intercrops should be so planned that the orchard is not irrigated when the trees are to be rested.

2.8 Canopy Management

There are two methods of training the pomegranate viz single stem and multi stem training. The young plants are properly trained to form a single stem with a number of well distributed scaffold limbs. Single stem is achieved by removing all the side shoots at the time of planting. On this main stem, 3-4 main branches are allowed to grow about 60-70 cm from the ground level. The second is multi stem system in which naturally arising stems are kept. The side branches are pruned and the tree is allowed to grow on a clean main stem. The fruits are borne terminally on short branches known as spurs. The older spurs which have lost the capacity of fruit bearing and also those shoots which interfere with others should be removed. It is desirable to encourage new growth on 1 to 3 years old wood.

2.9 Flowering and Fruiting

Flowering habit

The flowering habit of pomegranate is influenced by the prevailing climatic condition of the geographical region where it is grown (Pareek and Sharma 1993). In tropical climates, pomegranate flowers almost throughout the year whereas in subtropics, it flowers once a year. In areas where the temperature is low in winter, the tree is deciduous, but in tropical conditions, it is evergreen or partially deciduous. Evergreen cultivars of pomegranate flower throughout the year (Hayes, 1957). In southern India, the flowering season for evergreen cultivars was observed in three periods coinciding with June, October, and March (Nalawadi *et al.,* 1973). In the northern hemisphere, flowering occurs in April-May. However, flowering may continue until the end of summer, particularly in young trees. Such flowers are fertile, but the fruit will not mature properly because the plants enter dormancy during the cooler season in Mediterranean regions.

Flower bud development

Flower bud development takes place at varied times under tropical conditions. The time span between the start of flower bud elongation (growth) and anthesis ranges from 14 to 28 days depending on the variety and climatic conditions (Gur, 1986). Thus, in subtropical climate of the northern hemisphere, flowering occurs from the last week of March till the second week of May (Singh *et al.* 1978; Fouad *et al.,* 1979). Several distinct flushes of flowering on the same tree occur quite frequently. The time required for completion of flower bud development in Indian cultivars is between 20 and 27 days (Nalawadi *et al.,* 1973).

Flowering period

There are two flowering seasons in North India whereas Nalawadi *et al.* (1973) reported three flowering seasons in western India. In subtropical central and western India, there are 3 distinct seasons of flowering i.e. *ambe bahar* (January-February), *mrig bahar* (June-July) and *hasth bahar* (September-October). *Ambe bahar* is most commonly preferred by growers because of high yield consequent to profuse flowering compared to other flowering seasons (Singh *et al.* 1967; Prasanna Kumar 1998). In Karnataka, flowering was observed for 80-87 days in June- August (Nalawadi *et al.* 1973). In Punjab, only one flowering season was observed from April-June (Josan *et al.* 1979a). Under Delhi conditions, depending on the cultivar, flowering may occur once or twice a year (Nath and Randhawa 1959a). In the temperate climate of Himachal Pradesh, flowering takes place during the middle of April (Parmar and Kaushal 1982) and in Bihar, flowering occurs twice during February-March and July-August.

Inflorescence

The pomegranate inflorescence is a cyme (dichasial cyme). Flowers can appear solitary, in pairs or in clusters. The solitary flowers appear on spurs along the branches in most cases, while the clusters are terminal. Due to heavy drop of secondary and tertiary buds, the dichasial cyme appears to be solitary in clusters (Nath and Randhawa, 1959b). In the evergreen pomegranate cultivar, inflorescences of the spring flush are borne on mature wood of one-year-old shoots whereas the flowers which appear during July-August are borne on the current year's growth. In deciduous cultivars, flowers are borne on the current season's growth between July and August.

Kinds of Flowers

There are three kinds of flowers (Fig. 2.16) borne on the same pomegranate plant *viz*., staminate, hermaphrodite and intermediate which occur about 1 month after bud break on newly developed branches of the same year, mostly on spurs or short branches (Babu *et al.,* 2009b).

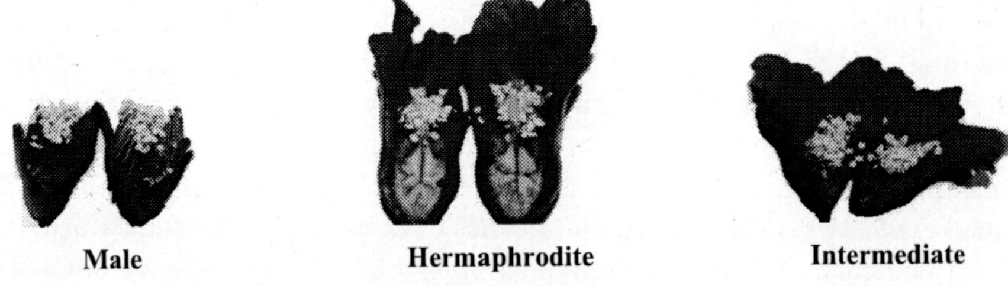

Fig. 2.16 : Male flower with rudimentary pistil; bisexual flower with long pistil; intermediate flower with medium pistil (See colour version on page 175)

The male flowers are campanulate (bell-shaped) whereas the bisexual flowers are urcerate (vase-shaped). The intermediate ones are tubular in shape. If fruit set takes place in such flowers, they may drop before reaching maturity, even if some fruits which reach maturity become misshaped. The ovary of the male flower is rudimentary whereas that of intermediate flowers are of the degenerating type. The bisexual flowers possess a well developed ovary with a broad base.

During the early balloon stage, the flower resembles a small pear with a greenish colour on its basal part and reddish colour on its apex or entirely dark red. As the flower matures (Fig. 2.17), it develops an orange-red to deep red sepal colour, which varies among different varieties. The petals are orange-red or pink and rarely white (Feng *et al.,* 1998; Wang 2003; Levin 2006; Beam Home 2007).

Fig. 2.17 : Matured flowers
(See colour version on page 175)

The sepals, 5-8 fused in their base, form a red fleshy vase shape. The sepals will not drop with fruit set but will stay as an integral part of the fruit as it matures, generating a fruit crowned with a prominent calyx. The flower has 5-8 petals. Their number usually equals the number of sepals. The petals, which alternate with the sepals, are separated and have a pink-orange to orange-red colour depending on the variety. The petals are obovate, very delicate, and slightly wrinkled. The multiple long stamens are inserted into the calyx walls in a circle and frequently number more than 300 per flower. They have an orange-red filament and yellow bilocular anthers that remain attached to the prominent calyx. Nectaries are located between the stamens and the ovary base (Fahan 1976). The carpels vary in number but are usually 8 superimposed in two whorls. They form a syncarpic ovary and are arranged in two layers.

Pollination

Pomegranate is both self- and cross-pollinated by insects, mainly bees. Wind pollination (anemophily) is reported to occur, but infrequently (Morton 1987). Emasculation and bagging studies on Indian, Turkmen, Israeli, and Tunisian pomegranate cultivars indicate that pomegranate flowers can self-pollinate and produce normal fruit (Nalawadi *et al.,* 1973; Karale *et al.,* 1993; Mars 2000; Levin 2006; Holland *et al.,* 2007). It was noted, however, that the degree of fruit set by self-pollination varies among different pomegranate cultivars (Levin 1978; Kumar *et al.,* 2004). In hermaphrodite flowers, 6 to 20% of pollen may be infertile while in male flowers, 14 to 28% are infertile. The size and fertility of the pollen varies with the cultivar and season (Morton, 1987). Yazici *et al.*, (2009) reported that higher levels of GA_3 affected barren flower formation in pomegranate.

Fruit Characteristics

The nearly round fruit is crowned by the prominent calyx. The apex of this crown is almost closed to widely opened, depending on the variety and on the stage of ripening. The fruit is connected to the tree with a short stalk.

Following fruit set, the colour of the sepals' skin in the developing fruit changes continuously from the prominent orange-red to green. In later stages of fruit maturation, the colour will change again until it reaches its final characteristic colour as the fruit ripens. The external colour ranges from yellow, green, or pink overlain with pink to deep red or indigo to fully red, pink or deep purple cover, depending on the variety and stage of ripening. There are some exceptional cultivars, such as the black pomegranate which acquires its black skin very early and remains black until ripening time. The skin (leathery exocarp) thickness varies among pomegranate cultivars.

Fig. 2.18 : A good fruit will have a high proportion of fleshy seeds in edible pulp

(See colour version on page 175)

The multi-ovule chambers (locules) are separated by membranous walls (septum) and fleshy mesocarp. The chambers are organized in a nonsymmetrical way. Usually the lower part of the fruit contains 2 to 3 chambers while its upper part has 6 to 9 chambers. The chambers are filled with many seeds (arils). The arils contain a juicy edible layer that develops entirely from outer epidermal cells of the seed, which elongate to a very large extent in a radial direction (Fahan 1976). The sap of these cells develops a turgor pressure that preserves the characteristic external shape of these cells. The colour of the edible juicy layer can vary from white to deep red, depending on the variety. Levin (2006) reports that occasionally metaxenia is observed such that there are several seeds of different colour within an individual pomegranate. The arils vary in size and the seeds vary in hardness among different varieties. Varieties known as seedless actually contain seeds that are soft.

There is no correlation between the outer skin colour of the rind and the colour of the arils. These colours could be very different or similar, depending on the variety. The external outer skin colour does not indicate the extent of ripening degree of the fruit or its readiness for consumption because it can attain its final colour long before the arils are fully ripened. The fruit ripens 5 to 8 months after fruit set, depending on the variety. The most pronounced difference in ripening time among cultivars is not derived from the differences in flowering dates but rather from the time required to ripening from anthesis.

2.10 Crop Rgulation

Pomegranate flowers continuously when watered regularly. The plants under such conditions may continue bearing flowers and bear small crop irregularly at different period of the year, which may not be desirable commercially. To avoid this trees are given bahar treatment. In this treatment, the irrigation is withheld two months prior to the bahar followed by light earthing up in the basin. This facilitates the shedding of leaves. The trees are then medium pruned 40-45 days after withholding irrigation. The recommended doses of fertilizers are applied immediately after pruning and irrigation is resumed. This leads to profuse flowering and fruiting. The fruits are ready for harvest 4-5 months after flowering. In tropical condition, there are three flowering seasons, viz., January-February *(ambia bahar)* June- July *(mrig bahar)* and September-October *(hasta bahar)*. The choice of flowering/fruiting is regulated taking into consideration the availability of irrigation water, market demand and pest/disease incidence in a given locality. The fruits of ambia bahar are ready for harvest in the month of June to August. As the fruit development takes place during dry months, they develop an attractive colour and quality thus suitable for exports. Similarly due to dry weather, the incidences of pest and disease attack are limited. However, ambia bahar can be taken only areas having assured irrigation facilities. The mrig bahar crop is harvested in the month of November to January. Usually this bahar is favoured as the flowering and fruiting period coincides with rainy season or immediately after rains, and the crop is taken without much irrigation. As the fruits develop during the rainy season and mature during winter, the colour and sweetness of the fruit is affected. The fruits from hasta bahar are harvested during the month of February to April. They have very attractive rind with dark coloured arils. Since the availability of the fruits during this season is limited, they fetch high value. Optimum water stress cannot be developed during this period as withholding of irrigation coincides with the rainy season. This leads to poor flowering and thus affects the yield.

Table 2.3 : Details of arrival pattern of pomegranate according to *Bahar* treatment

S. No.	Bahar	Flowering Time	Period of Harvest	Remarks
1.	Ambe	Jan.-Feb.	June–August	More flowering and high yield.
2.	Mrig	June-July	November-January	More prone to diseases and insect pests; fruit quality is not very good; it should be avoided in areas receiving more rainfall during July-September.
3.	Hasta	September-October	February-April	Flowering good; fruit colour and quality best; less diseases and insect pests as fruit mature in cold season; fetches higher market price; preferred for export.

Table 2.4 : Advisory for *Ambe bahar* crop

S.No.	Days	Stage	Period	Horticultural Operations
Pre- Ambe bahar Horticultural Operations				
1.	-	Crop growth	June 15-30	• Moderate to heavy pruning to remove disease affected broken, criss-cross branches, water sprouts, suckers and opening canopy to improve light penetration. Apply Bordeaux paste (10%) on cut ends. • Ploughing in inter- space. • Bronopol (0.5 g/l) + Coper oxicloride (COC) (2.5g/l)/ Kocide (2g/l) + Spreader and sticker (0.5g/l) • Apply recommended dose of manure and Fertilizer as per Table.1 along with Neem cake 0.5 kg+ vermicompost 1.5 kg+ calcium ammonium nitrate 50g/ tree in circular trench (15-20 cm wide of 8-10 cm depth) at 45-60cm apart from the stem, cover the trenches properly with soil and start irrigation.
2.	-	Crop growth	July 1-15	• Spray Bordeaux mixture (1%) or Kocide (2g/l)
3.	-	Crop growth	July 16-31	• Spray Bronopol (0.5 g/l)+ COC (2.5g/ l)/Kocide (2g/l)+ Spreader and sticker (0.5g/l) • Weeding.
4.	-	Crop growth	Aug. 1-15	• Spray Bordeaux mixture (1%) or Kocide (2g/l) • Remove water sprouts and suckers.
5.	-	Crop growth	Aug. 16-31	• Evening spray Bronopol (0.5 g/l) + Carbendazim 50WP (1g/l) + Acetamiprid 20SP@ 0.3g/l +spreader and sticker (0.5ml/l). • Harrowing and Weeding
6.	-	Crop growth	Sept. 1-30	• Spray Bordeaux mixture 1% or Kocide (2g/l) • Weeding
7.	-	Stress	Oct. 1-30	• Remove water sprouts and suckers. • Ploughing / Harrowing in inter space. • Spray Bronopol (0.5 g/l) + COC (2.5g/ l)/Kocide (2g/l)+ Spreader and sticker (0.5g/l)

8.	-	Stress	Nov. 1-15	• Spray Bordeaux mixture 1% or Kocide (2g/l)
9.	-	Stress	Nov. 16-30	• Spray Bronopol (0.5 g/l) + COC (2.5g/l)/Kocide (2g/1) + Spreader and sticker (0.5g/l)

Horticultural Operations in ambe bahar

1.	0-7	Defoliation	Dec.1-7	• Spray 1% Bordeaux mixture 2 days before defoliation • Defoliate with Ethrel (1.5-2 ml/l) + DAP 5g/l.
2.	8-15	leaf fall (85-100%), emergence of new leaves and sprouts	Dec. 8-15	• Remove weeds and suckers • Collect & burn fallen leaves /debris from the orchard • Harrowing in interspaces. • Apply full dose of well rotten FYM and P, 1/3rd N&K fertilizers (Table. 2). + Micronutrients ($ZnSO_4$, $FeSO_4$, $MnSO_4$ each 25g and 10g Borax (Boron) /tree)+ Neem Cake 1-1.5 kg / tree + Vermicompost 2 kg/tree+ Phorate 10G @ 25g/tree or Carbofuran 3G @ 40g/tree in shallow trench or ring (15-20 cm wide of 8-10cm depth) at 45-60cm apart from the stem, cover the trenches properly with soil and give light irrigation immediately after fertilizer application.
3.	16-31	New flush emergence/ Flower bud initiation	Dec. 16-31	• Evening spray Streptocycline (0.5g/l) + Carbendazim 50WP (1g/l) + Thiamethoxam 25WG @ (0.3g/l) + spreader and sticker (0.5ml/l) • Spray ZnSO4,(3g/l) + MnSO4 (6g/l)
4.	32-46	Flower bud initiation continue	Jan. 1-15	• Morning spray Salicylic acid formulation @ 0.3g/1 • Spray Boric acid (2g/l) or Solubor (2g/l) • Evening spray Bronopol (0.5g/l) + Ziram 80% WP (2g/l) + Acetamiprid 20SP@ (0.3g/l) + Spreader and sticker (0.5ml/l) • Apply 1/3rd dose of Niotrogen + 1/3rd dose of potash
5.	47-62	Flower bud development	Jan. 16-31	• Evening spray Streptocycline (0.5g/l) + Thiophanate Methyl 70 WP (1g/l) + Thiamethoxam 25WG @ (0.3g/l) + spreader and sticker (0.5ml/l)

6.	63-77	Flowering	Feb. 1-15	• Morning spray salicylic acid formulation @0.3g/1 • Evening spray Captan 50WP (2.5g/l) + Bronopol (0.5g/l) + Spreader and sticker (0.5 ml/l) + Methomy l 40% SP @ (1g/l) • Remove suckers & weeding
7.	78-90	Flowering continue & fruit setting	Feb 16-Feb 28	• Spray $ZnSO_4$ (3g/l) + $MnSO_4$ (6g/l) • Spray Streptocycline (0.5g/l) + Mancozeb 75% WP (2g/l)+ Cypermethrin 25% EC (1ml/1) + Spreader and sticker (0.5ml/l)· Apply 1/3rd dose of Nitrogen + 1/3rd dose of potash
8.	91-106	Flowering continue & Fruit setting	March 1-15	• Evening spray Bronopol (0.5g/l) + Carbendazim 50WP (1g/l) + Spreader and sticker (0.5ml/l) • Morning spray Solubor (2g/l) + commercial micronutrient mixture (1g/l)
9.	107-122	Fruit setting continue	March 16-31	• Morning spray salicylic acid formulation @0.3g/1 • Evening spray Steptocycline (0.5g/l) + Thiophanate Methyl 70WP (1g/l) + Cypermethrin 25% EC (1ml/l) + Neem Seed Kernel Extract @50g/l (75g if entire seed is used) + spreader and sticker (0.5ml/l) • Need based fruit thinning • Remove suckers & do weeding
10.	123-138	Fruit enlargement	Apr 1-15	• Morning spray Magnesium sulphate (2g/l) + $ZnSO_4$ (3g/1) + $MnSO_4$ (6g/1) • Evening spray Bronopol (0.5g/l) + Fosetyl Al 80% WP (2g/l) + spreader and sticker (0.5 ml/l) + Thiamethaxam (0.3g/l)
11.	139-153	Fruit enlarge-ment continue	Apr 16-30	• Morning spray salicylic acid formulation @ 0.3g/l • Evening spray Bordeaux mixture (0.5g/l)
12.	154-169	Fruit	May 1-15	• Morning spray calcium nitrate (15g/l) • Evening spray Bronopol (0.5g/l) + (0.5g/l) + Mancozeb 75% WP

13.		Enlargement continue		(2g/l) + spreader and sticker (0.5ml/l) + Deltamethrin 2.8Ec@1.5ml/l • Remove suckers & do weeding
14.	170-185	Fruit maturation	May 16-31	• Spray Streptocycline (0.5g/l) + Propiconazole (1ml/l) + Neem seed kernel extract @50g/l (75g if entire seed is used) +spreader and sticker (0.5ml/l)
15.	186-200	Fruit maturation continue	June 1-15	• Evening spray Bronopol (0.5g/l) + Neem seed kernel extract @50g/l (75g if entire seed is used) + spreader and sticker (0.5ml/l)
16.	201-215	Fruit ripening	June 16-30	• Spray Potassium nitrate@10g/l or 0:0:50 @10g/l
17.	216-230	Fruit ripening/ harvesting	July 1-15	• Harvest

Note: Streptocycline is streptomycin sulphate 90% + oxy Streptocycline 10%; Bronopol is 2-bromo, 2-nitro propane, 3-diol.

Precautions/Suggestions

1. Streptocycline and Bronopol should be used only if orchard is having BBD symptoms.
2. Spraying should be done either in the morning or evening hours.
3. Need based spraying of fungicides, bactericides and insecticides should be done to minimise residues in fruits.
4. Irrigation interval may vary from 1-3 days depending upon weather condition, age of plant, crop stage, soil type etc.
5. Provide less irrigation during new leaf emergence and flower development stages as excess water during these stages may lead poor flowering.
6. There should not be moisture stress at fruit setting, fruit maturation and fruit ripening.
7. Always prepare Bordeaux mixture fresh and use on the same day.
8. In general, the spray schedule for BBD (Bacterial Blight Diseases) varies from 7-20 days upon weather condition, crop stage and its level of infection on plant parts.
9. Judicious irrigations schedule should be followed to avoid wilt problem.
10. Number of spraying schedules may be increased or decreased as per actual need.

11. Use good quality and non –ionic spreader sticker while spraying. However, no stickers should be used with Bordeaux mixture.
12. Combine insecticides, fungicides or micronutrient sprays with bactericidal sprays depending on compatibility to reduce number of sprays. Mixture should not form precipitate.
13. Integrated nutrient management practices using different sources of FYM, poultry manure, vermicompost, goat manure, neem cake etc. are must to keep the orchard healthy for developing natural immunity to various diseases / pests.

Table 2.5 : Advisory for *Mrig Bahar* crop

S. No.	Days after defoliation	Stage	Period	Pesticides
Sprays During Crop Period				
1.	0-7	Defoliation	May 16-22	Spray 1% Bordeaux mixture before defoliation
2.	8-14	85-100 % leaf fall	May 23-29	• Remove fallen leaves and debris from the orchard and burn • Drench soil with bleaching powder (33% Cl)@25Kg/1000 liters/1 ha • Spray Copper Oxychloride 50WP (2.5g/l) + Bronopol (0.5g/l) + +spreader and sticker (0.5 ml/l)
3.	15-21	First flush of leaves	May 30-June 5	• Morning spray salicylic acid formulation @ a.i.0.3g/l • Evening spray Streptocycline (0.5g/l) + Carbendazim 50WP (1g/l) + Thiamethoxam 25WG @ 0.3g/l+spreader and sticker (0.5ml/l)
4.	22-28	Flower initiation	June 6-12	Bronopol (0.5g/l) + Ziram 80% WP 2g/l +spreader and sticker (0.5ml/l)
5.	29-35	Flowering	June 13-19	• Morning spray mixture of zinc sulphate (3g/l), Solubor (2g/l), chelated iron (3g/l) • Evening spray Streptocycline (0.5g/l) + Carbendazim 50WP (1g/l) + Acetamiprid 20SP@ 0.3g/l/ +spreader and sticker (0.5ml/l)

6.	36-42	Flowering	June 20-26	Spray Captan 50WP (2.5g/l) + Bronopol (0.5g/l) + spreader and sticker (0.5ml/l)
7.	43-49	Flowering 100%	June 27-July 3	Spray Streptocycline (0.5g/l) + Mancozeb 75% WP (2g/l) + Imidacloprid (0.5ml/l) 17.8 SL @ 0.3ml/l+spreader and sticker
8.	50-56	Fruit set starts	July 4-10	• Morning spray salicylic acid formulation @ a.i.0.3g/l • Evening spray Streptocycline 0.5g/l) + Ziram 80% WP (2g/l) +spreader and sticker (0.5ml/l)
9.	57-63	Fruit setting	July 11-17	• Morning spray Solubor 2 g/lit + commercial micronutrient mixture 1g/l • Evening spray Steptocycline (0.5g/l)+Thiophanate Methyl 70WP (1g/l) + Cypermethrin 25%EC (1 ml/l) + Neem Seed Kernel Extract @50g/l (75g if entire seed is used). in evening + spreader and sticker (0.5ml/l).
10.	64-70	Fruit setting	July 18-24	• Morning spray Magnesium sulphate (2g/l) • Evening spray Bronopol (0.5g/l) + Fosetyl Al 80% WP (2g/l) + spreader and sticker (0.5ml/l)
11.	71-77	Fruit enlargement	July 25-31	• Morning spray salicylic acid formulation @ a.i.0.3g/l • Evening spray Bordeaux mixture (0.5%)
12.	78-84	Fruit enlargement	Aug. 1-7	• Morning spray calcium nitrate (15 g/l) • Evening spray Bronopol (0.5g/l) (0.5g/l) + Mancozeb 75% WP (2g/l) +spreader and sticker (0.5ml/l)
13.	85-91	Fruit enlargement	Aug. 8-15	Spray Streptocycline (0.5g/l) + Carbendazim 50 WP (1g/l) + methomyl 40% SP @ 1g/l +Neem seed kernel extract @50g/l (75g if entire seed is used) + spreader and sticker (0.5ml/l)
14.	92-98	Fruit enlargement	Aug. 16-21	• Morning spray salicylic acid formulation @ a.i.0.3g/l

				• Evening spray Bronopol (0.5g/l) (0.5g/l) + Ziram 80% Wp (2g/l) + Azadirachtin 10000 ppm (3ml/l) +spreader and sticker (0.5ml/l)
15.	99-105	Fruit enlargement	Aug. 22-28	Spray Bordeaux mixture (0.5%)
16.	106-112	Fruit enlargement	Aug. 29 - Sep. 4	Spray Steptocycline (0.5g/l) (0.5g/l) + Mancozeb 75% WP (2g/l) + spreader and sticker (0.5ml/l)
17.	113-119	Fruit enlargement	Sep. 5-11	• Spray Captan 50WP(2.5g/l) + Bronopol (0.5g/l)+ Methomyl 40%SP@ 1g/l +spreader and sticker (0.5ml/l) • Drench with bleaching powder (33% Cl) @25Kg/1000 liters/1 ha+spreader and sticker (0.5ml/l)
18.	120-126	Fruit enlargement	Sep.12-18	• Morning spray Calcium nitrate (8-10 g/l) • Evening spray Steptocycline (0.5g/l) (0.5g/l) + copper hydroxide 77WP (2g/l) 77% WP (2g/l) +spreader and sticker (0.5ml/l) in evening same day
19.	127-133	Fruit enlargement+ Aril colour developmentl)	Sep.19-25	Spray Bronopol (0.5g/l) + Thiophanate Methyl 70WP (1g +Acetamiprid 20SP @ 0.3g/l+spreader and sticker (0.5ml/l)
20.	134-140	Fruit enlargement+ Aril colour development	Sep. 26-Oct. 2	Spray Steptocycline (0.5g/l) (0.5g/l) + Mancozeb 75% WP (2g/l) +spreader and sticker (0.5ml/l)
21.	141-147	Fruit enlargement and development		Oct. 3-9 Spray Bordeaux mixture (0.5%)
22.	148-154	Fruit enlargement development		Oct. 10-16 Spray Bronopol and (0.5g l) + Captan 50% WP (2.5g/l) + spreader and sticker (0.5ml/l)
23.	155-161	Fruit enlargement and development	Oct. 17-23	Spray Steptocycline (0.5g/l) (0.5g/l) + Mancozeb 75% WP (2g/l) + Lambda cyhalothrin 5EC/CS (0.5g/l) + spreader and sticker (0.5ml/l)

24.	162-168	Fruit enlargement and development	Oct.24-30	Bordeaux mixture (0.5%)
25.	169-184	Fruit enlargement & development	Oct.31-Nov14	spray Potassium dihydrogen phosphate @10g/l + Spray Bronopol (0.5g/l) + Captan 50% WP (2.5g/l) + spreader and sticker (0.5ml/l)
26.	185-199	Fruit Maturity	Nov.15- 29	Spray Steptocycline (0.5g/l) + copper hydroxide 77WP (2g/l)+ Neem seed kernel extract @50g/ l (75g if entire seed is used) or Azadirachtin 1500ppm @ 3ml/ l+spreader and sticker (0.5ml/l)
27.	200-214	Fruit Maturity (1 month before harvest)	Nov. 30-Dec.14	Spray Potassium nitrate@10g/l or 0:0:50 @10g/ lBordeaux mixture @ 0.5% only under adverse weather conditions
28.	215-230	Fruit ripening	Dec. 15-31	Harvest
Sprays During Rest Period				
29.	-	Rest	Jan. 1-15	Apply Bordeaux paste (10%) on pruned ends Immediately after pruning spray Bordeaux Mixture (1%)
30.	-	Rest	Jan. 16-31	Spray Streptocycline or Bronopol @0.5g/l + Copper Oxychloride 50WP (2.5g/l) + spreader and sticker (0.5ml/l)
31.	-	Rest	Feb. 1-15	Spray Bordeaux Mixture (1%)
32.	-	Rest	Feb. 16-28	Spray Streptocycline or Bronopol @0.5g/l+ copper hydroxide 77WP (2g/l)+ spreader and sticker (0.5ml/l)
33.	-	Rest	Mar. 1-15	Spray Bordeaux Mixture (1%)
34.	-	Rest	Mar.16- 31	Spray Streptocycline or Bronopol @0.5g/l + Copper Oxychloride 50WP (2.5g/l) +spreader and sticker (0.5ml/l)
35.	-	Stress	Apr.1-15	Spray Bordeaux Mixture (1%)
36.	-	Stress	Apr.16- 30	Spray Streptocycline or Bronopol @0.5g/l + copper hydroxide 77WP (2g/l) + spreader and sticker (0.5ml/l)
37.	-	Stress	May 1-15	Spray Bordeaux Mixture (1%)

Note:

1. To prepare spray mixture, prepare dilute solutions of each chemical separately and mix to make total volume. If precipitate is formed, either mixture chemicals are not compatible or pH is not proper.
2. In case no rains are there for long duration or blight is not increasing, sprays can be taken at 10-15 days interval instead of 7 days.
3. Bronopol is -2-bromo-2-nitro-1, 3-diol.

Precautions

- Take only need based sprays at recommended doses, too many sprays increase the disease.
- Always remove and burn all affected fruits before starting any spray.
- Combine insecticides, fungicides or micronutrient sprays with bactericidal sprays depending on compatibility to reduce number of sprays. Mixture should not form precipitate.
- Take without fail, additional spray with a bactericide after the rains -when plant surfaces dry up.
- Always (rains or no rains) mix good quality non-ionic spreader sticker with sprays. Do not use spreader sticker with bordeaux mixture.
- Always prepare Bordeaux mixture fresh and use on the same day

Table 2.6 : Cultural operations for *Mrig Bahar* crop

S. No.	Days after defoliation	Stage	Period	Operations
1.	0-7	Defoliation	May 16-22	• Defoliate with Ethrel (1.5-2ml/l)+DAP5g/l • Remove weeds and suckers· Apply 2/3rd dose of FYM + Micronutrients (Zn, Fe, Mn, Boron each 25 g per plant) + Neem Cake 1-2 Kg/plant + Vermicompost 1 Kg/plant + Carbofuran 3G@40g/plant in wet soil in a ring around the plant • Apply 1/4th dose of nitrogen + 1/3rd dose of phosphorus and 1/3rd dose of potash

				• Give light irrigation immediately after fertilizer application
2.	8-14	85-100 % leaf fall	May 23-29	• Remove fallen leaves and debris from the orchard and burn • Drench soil with bleaching powder (33% Cl) @25Kg/1000 liters/1 ha
3.	15-21	First flush of leaves	May 30-June 5	Irrigate
4.	22-28	Flower initiation	June 6-12	Apply 1/4th dose of nitrogen Fertigate N; P; K:12:61:00 @ 8 kg/ha/application.
5.	29-49	100% Flowering	June 13-26	Give 15 applications on alternate days for a month through irrigation. Irrigate
6.	50-63	Fruit set starts	July 4-17	Apply 1/4th dose of nitrogen in soil Fertigate N;P;K::19:19:19 @ 8 kg/ha/application. Give 15 applications on alternate days for a month through irrigation
7.	64-70	Fruit setting	July 18-24	Remove weeds Irrigate regularly
8.	71-126	Fruit set 100% Fruit enlargement	July 25-Sep.18	Fertigate N;P;K::00:52:34 Mono-Potassium Phosphate @ 2.5 kg/ha/application -Give 15 applications on alternate days for a month through irrigationApply 1/4th dose of nitrogen in soil in August first week Irrigate
9.	127-140	Fruit enlargement+ Aril colour development	Sep.19-Oct 2	• Remove weeds and suckers • Irrigate
10.	141-184	Fruit enlargement & development	Oct. 3-Nov.14	
11.	185-199	Fruit Maturity	Nov. 15-29	
12.	200-214	Fruit Maturity 1 month before harvest	Nov. 30-Dec.14	Fertigate Calcium Nitrate @12.5 kg/ha/application give 2 applications at 15 days interval Give moderate Irrigation
13.	215-230	Fruit ripening	Dec. 15-31	Harvest mature fruits

Operations During Rest Period				
14.	-	Rest	Jan. 1-15	Do heavy pruning to remove blight affected and old dry branches. Apply 1/3rd dose of FYM, Neem cake @ 1Kg /plant and 2/3rd dose each of phosphorus and potash Light Irrigation
15.	-	Rest	Jan. 16-March 31	Light Irrigation
16.	-	Stress	Apr. 1- May 15	Stop Irrigation

Note:

1. Irrigate depending on soil type, plant age and stage and weather conditions. Apply nitrogen through CAN in red soil / Urea in black soil; Phosphorus through DAP and SSP and Potash through MOP depending on age of plant.

3

Integrated Disease and Insect Pest Management

3.1 Integrated Insect Pests Management

1. Anar butterfly [*Deudorix (Virachola) isocrates* Fabricus]

Fig. 3.1 : Fruit borer damaged fruit
(See colour version on page 176)

Pomegranate fruit borer or pomegranate butterfly is the most widespread, polyphagous and destructive pest distributed all over India and common in Asia. Peak incidence is during the month of August in monsoon season, while it is more during November/December in winter crop. Infestation from flowering to button stage causing loss up to 50 per cent of the fruit. Adult bluish brown butterfly, female with V shaped patch on forewing. Full-grown larvae are dark brown with short hair and white patches all over the body and measures about 16 to 20mm long. The larvae bore into the pomegranate fruits soon after hatching. Once inside the fruit, larvae (approx 2cm length) feed on the flesh and seeds. The bored hole is plugged by the last abdominal segment of the larva. When fully grown, the larva comes out by boring through the hard shell and spins a web, which ties the fruit, stalk to the main branch. Offensive smell and excreta of caterpillars coming out of the entry holes with excreta stuck around the holes. The fruits rot and drop off. The holes ultimately expose the rest of the fruit to disease, and typically rot off the tree.

Natural enemies of anar butterfly/pomegranate fruit borer

Predators: Lacewing, ladybird beetle, spider, red ant, dragon fly, robber fly, reduviid bug, praying mantis, black drongo (King crow), wasp, common mynah, big eyed bug (*Geocoris* sp), earwig, ground beetle, pentatomid bug (*Eocanthecona furcellata*) etc.

Management

Cultural control

- Remove eggs from calyx.
- Collect and destroy damaged fruits
- Follow clean cultivation as weed plants serve as alternate hosts

Mechanical control

- Prune the affected parts of the plant and destroy.
- Detect early infestation by periodically looking for drying branches.
- Use light trap @ 1/ acre to monitor the activity of adults

Biological control

- Release *Trichogramma chilonis* @ one lakh/acre.

Chemical control

- Spray Deltamethrin @ 0.0028% at the time when more than 50% of fruits have set.
- Repeat after two weeks with Fenvalerate @ 0.005% in non-rainy season Quinalphos @ 0.06% is also effective. The number of sprays depends on severity of infestation.

2. Shot Hole Borer (*Xyleborus perforans Wollastan*)

Shot hole Borer eggs are oval or round, shiny and iridescent white. The larvae are white and legless, and can be up to 4 mm long. The adult is about 2-3 mm long, black to reddish-brown, and cigar shaped. They have a short, stubbed head capsule, with chewing mouthparts. Male adults do not fly. Adult females bore into the basal part of the stem and roots causes Small shot holes on roots, main trunk, wilting and finally leads to death of the tree.

Natural enemies of shot hole borer

Parasitoids: *Trichogramma* sp, *Tetrastichus* spp., *Telenomus* spp., *Chelonus blackburni, Carcelia* spp. *Campoletis chlorideae, Bracon* spp., etc.

Predators: *Chrysoperla* spp., rove beetle, spider, parasitic wasp, coccinellids, robber fly, dragonfly, big eyed bug (*Geocoris* sp.), praying mantis, red ants, pentatomid bug (*Eocanthecona fucellata*), earwigs, ground beetles, common mynah, king crow etc.

Management

Cultural control

- Avoid water logging and rake the soil.

Mechanical control

- Infested trees should be uprooted and brunt, especially the root zone.
- Prune the affected fruits and buds of the trees and destroy.
- Detect early infestation by periodically looking for drying branches.
- Infested young plants should be uprooted and burnt.

Chemical Control

- Apply Chlorpyriphos (0.1%). The insecticides should be applied as pastes to the tree trunks with 40% geru (red soil) and Copper Oxychloride at 0.25%. Addition of geru and Copper Oxychloride to the insecticides increased the control of insect pests.

3. Stem Borer (*Coelosterna spinator* Febricius)

Life Cycle

Egg: Eggs are laid in young living plants in stems and is deposited under the bark. The number of eggs laid by female is 20-40.

Grub: Newly emerged larva is about 1/4 of an inch long, the mature larva is about 2.1/2 inches long. On hatching the larva feeds on the soft tissues around the oviposition cavity and then bores into the stem and roots. The length of the larval period is about 9-10 months.

Pupa: Period is 16 to 18 days.

Adult: Pale yellowish-brown body with light grey elytra and are 30 to 35 mm long.

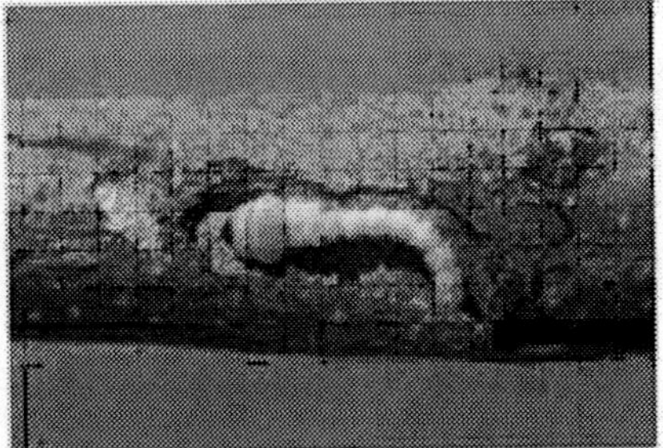

Fig. 3.2 : Larva bores into the stem
(See colour version on page 176)

The beetle emerges by eating a circular hole through the bark. Adult beetles are 1.1/4 to 1.1/2 inches long, dull, yellowish-brown, the sides of the body and legs bluish, elytra yellowish-grey with a large number of black spots varying in size from a pin's head to minute specks. There is only one generation per year and longevity of beetles is 45 to 60 days.

Damage symptoms

- The grubs bore inside the trunk and feed on sapwood.
- Adult beetles are active by the day and feed by gnawing the green bark of shoots.
- Holes on bark of main stems, excreta and dry powdered material are usually seen near the base of plants.

Natural enemies of stem borer

Predators: Damsel bugs, elm leaf beetle, spiders, tachinid flies, big eyed bugs (*Geocoris* sp), braconid wasp etc.

Management

Cultural control

- Clean cultivation: For healthy growth keep basin clean.
- Soil health: Avoid water logging, keep soil raked and aerated, to reduce invasion of shot hole borer.
- Detection of pest: The infestation should be detected periodically by looking out for drying branches.
- Moderate to heavy pruning to remove disease affected, broken, crisscross branches, water sprouts, suckers and opening canopy to improve light penetration.

Mechanical control

- Uproot infested trees and burn.

4. Leaf Eating Caterpillar (*Achea janata*)

Moth are stout, pale reddish brown with wavy lines on fore wings and black hind wings having a medial white bank and large white spots on the out margin, anal and apical margins of the wings are fringed with hair. A monkey face-like appears on the dorsum of thorax. The caterpillars are semiloopers with various colours some are grey with red / brown lateral stripes while others are bluish grey specked with blue-black and having yellowish lateral stripes. Pupae are glistening dark green, later becoming brown. Caterpillar feeds voraciously on leaves, tender petioles and young capsules. Adults suck the juice from fruits resulting complete defoliation of plants leaving only veins of the leaves.

Management

- Collect and destroy the caterpillars mechanically.
- Spray Chlorpyriphos (0.02%)

5. Bark-Eating Caterpillar *(Inderbela telraonis)*

The caterpillar bores the bark and feeds inside. Several holes can be seen on the trunk and the trees loose productivity. Wood dust and faecal matter hanging in the form of a web around the affected portion is indication of the borer activity.

Management

- The webs around the affected portion should be cleaned.
- Cotton swab soaked in petrol or kerosene should be inserted in the holes and sealed with mud.
- Sprays with Quinalphos (2 ml/litre of water) on tree trunk is effective in controlling the pest.

Parasitoids

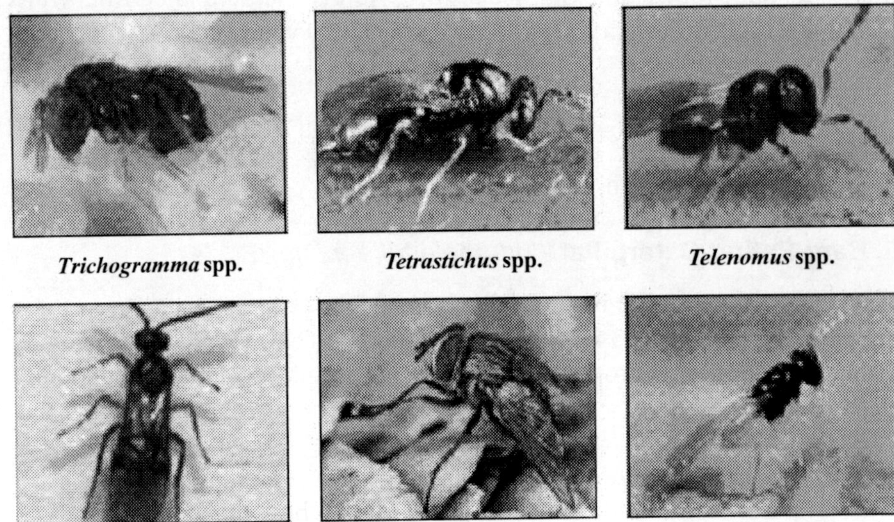

Trichogramma spp. | *Tetrastichus* spp. | *Telenomus* spp.

Bracon spp. | *Carcelia* spp. | *Encarsia inaron*

Predators

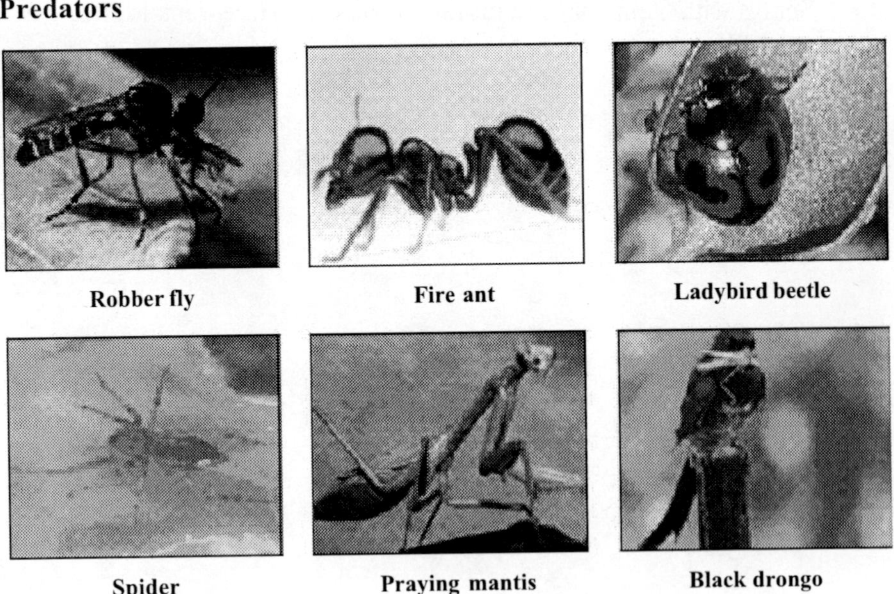

Robber fly | Fire ant | Ladybird beetle

Spider | Praying mantis | Black drongo

Fig. 3.3 : Important natural enemies of pomegranate insect pests

(See colour version on page 176)

Fig. 3.4 : Plants suitable for ecological engineering in pomegranate orchard
(See colour version on page 177)

6. Whitefly (*Siphorinus Phillyreae* Haliday)

Fig. 3.5 : White flies on leaves
(See colour version on page 178)

Adult females lay eggs on the lower surface of apical leaves often in circles or small groups. Eggs hatch after a week. The crawlers dig their mouth parts into the leaf tissue for sucking the sap and remain static as "scales" throughout the remaining part of their larval and pupal period. Serious damage is caused by the excretion of honeydew secreted by the by whitefly, which runs down to the fruit and the upper surface of leaves. Under moist conditions, sooty molds can develop on the honeydew, reducing photosynthesis and hindering respiration of plants. The damage by whitefly also leads to yellowing of leaves and stunted growth, in severe cases leading to shedding of leaves.

Natural enemies of whitefly

Parasitoids: *Encarsia inaron, Eretmocerus* sp, *Chrysocharis pentheus* etc.

Predators: *Cryptolaemus montrouzieri,* lacewings, ladybird beetles etc.

Management

Cultural control

- Field sanitation.
- Removal of host plants.
- Maintain adequate aeration by proper training and pruning.

Mechanical control

- White flies can be trapped by hanging bright yellow sticky traps coated with polybutene adhesive at the height of the crop canopy.
- Spraying water with high volume sprayer by focusing the nozzle towards the under surface of leaves helps in washing out the honeydew, eggs, larvae, pupae and adult whitefly.

Chemical control

- Spray Dimethoate (0.045%) at the initiation of infestation and repeat once after fortnight.
- Spray Cyantraniliprole 10.26% OD @ 360 ml in 400 l of water / acre.

7. Aphids (*Aphis punicae* Passerini)

Aphids are yellowish green in colour. They suck the cell sap from the lower surface of the leaves and devitalize the plant. They secrete sweet sticky substance, which attracts fungal growth. The affected leaves show chlorotic patches. High humidity favours the multiplication of aphids.

Fig. 3.6 Aphids on flower and shoot
(See colour version on page 178)

Natural enemies of aphids

Parasitoids: *Aphidius* sp, *Aphelinus* sp etc.

Predators: Parasitic wasp, *Chrysoperla* spp., ladybird beetle, predatory mite, syrphid fly etc.

Management

Cultural control

- Collect and destroy the damaged plant parts.
- Maintain adequate aeration by proper training and pruning.

Biological control

- Release first instar larva of *Chrysoperla zastrowi sillemi* @ 15 / flowering branch (four times) at 10 days interval from flower initiation during April.

Chemical control

- Give single spray of Neem Seed Kernel Extract (2%) at the initiation of new flush.
- Spraying with Dimethoate (0.03%) at 15 days interval effectively controls the aphid population.
- Spray Cyantraniliprole 10.26% OD @ 360 ml in 400 l of water / acre.

8. Mealybugs (*Ferrisia virgata* Cockerell)

Adult females are oval with waxy filaments all over the body. Nymphs and adults of mealy bugs suck sap from the leaves and tender shoots. Leaves show characteristic curling symptoms similar to that of a virus. A heavy black sooty mould may develop on the honeydew like droplets secreted by mealy bugs. The infestation may lead to fruit drop. The bugs lay eggs into the soil remain dormant till the next bahar. The nymphs hatch from the eggs during the next bahar and attack the plants.

Fig. 3.7 : Mealybug infestation on fruit

Natural enemies of mealybug

Predators: *Menochilus sexmaculatus, Rodolia fumida, Cryptolaemus montrouzieri etc.*

Management

Cultural control

- Collect and destroy the infested plant parts.
- Remove alternate hosts.

Mechanical control

- The plants in the vicinity of the vineyard serving as alternate hosts for the mealy bugs should be destroyed.
- Pasting a grease band of 5cm width on the main stem prevents the crawlers from reaching the bunch. Unlike the adults, the crawlers are free from waxy coating and therefore the crawler stage is the most effective for spraying pesticides.

Biological control

- Release *Cryptolaemus montrouzieri* near the site of mealybug @ 10/tree.

Chemical control

- Spraying of insecticides like Dichlorvos (0.02%) or Malathion (0.2%) with fish oil rosin soap was found to control the insect population.

9. Scale insects

The scale insects can be identified by presence of small black swollen spots on the branches sand the fruits. Adults and pupa suck the cell sap from the fruit and tender shoots causing drying of branches. In case of severe infestation, the whole tree dries up. The insects secret honey dew like substance which attracts black sooty mould. As a result, all the leaves and the branches turn blackish affecting the growth of the plant.

Fig. 3.8 : Black Scale on fruit
(See colour version on page 178)

Management

- Removal and destruction of alternate hosts, which harbour the scale insects.
- Spraying the affected patches with Dimethoate (0.045%) or Quinalphos (0.06%) at 15 days interval helps to control the pest.

10. Fruit fly (*Drosophila*)

The attack is prominent during the rainy season. The female lays eggs under the rind of the fruits by puncturing. After hatching the caterpillars feed on the pulp. The affected fruits cease to develop and drop. During the rainy season, water enters through the small holes created by the females leading to fruit rot. The damage leads to severe economic losses.

Management

- Since the pest remains inside the fruit chemical control measures are ineffective.
- Using 'fly traps' containing Methyl Eugenol and an insecticide can control the pest.

11. Thrips (*Scirtothrips Dorsalis* Hood or *Rhipiphorothrips cruentatus* Hood)

The adults ***(Rhiphiphorothrips cruentatus)*** are minute, slender, soft bodied insects with heavily fringed wings, blackish brown with yellowish wings and measure 1.4 mm long. Scirtothrips dorsalis adult is straw yellow in colour. The nymphs are minute yellowish brown in colour. Female lays on an average 50 dirty white bean-shaped eggs on the under surface of leaves. The incubation period is 3-8 days. Newly hatched nymphs are reddish and turn yellowish brown as they grow. Pupal period lasts 2-5 days. Both nymphs and adults feed on the underside of the leaves by rasping the surface and sucking the oozing cell-sap causing leaf tip turn brown and get curled, drying and shedding of flowers and scab on fruits which will reduce the market value.

Fig. 3.9 : Thrips affected fruit

Fig. 3.10: Thrips affected leaves

(See colour version on page 178)

Natural enemies of thrips

Parasitoid: *Ceranisus menes, etc.*

Predators: Syrphid fly, minute pirate bug, praying mantis, predatory thrips, damsel bug, lacewing, coccinellid, spider *etc.*

Management

Cultural control

- Keep basin clean.
- Maintain adequate aeration by proper training and pruning.

Mechanical control

- Prune the affected parts of the plant and destroy.
- Detect early infestation by periodical monitoring for drying branches.

Chemical control

- Spray Dimethoate (0.06%) prior to flowering is important and in severe condition spray Fenitrothion (0.05%) and repeat after fruit set.
- Spray Cyantraniliprole 10.26% OD @ 300 ml in 400 l of water / acre.

12. Mites (*Oligonychus punicae, Aceria granati*)

Adult and nymphs feed on the lower leaf surface, causing brown patches, curling and drying.

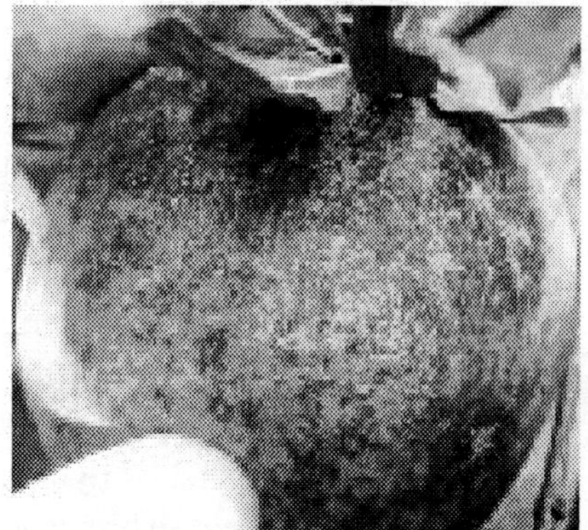

Fig. 3.11 : Mite damaged fruit (See colour version on page 178)

Management

1. Spray Dicofol (0.04% once or twice depending upon severity.

3.2 Integrated Disease Management

1. Bacterial Blight (*Xanthomonas axonopodis* pv *punicae*)

Fig. 3.12 : Bacterial blight affected fruits

Bacterial blight of pomegranate affects leaves, twigs, and fruits. Infected fruit and twigs are potential sources of primary inoculum. The secondary spread of bacterium is mainly through rain and spray splashes, irrigation water, pruning tools, humans, and insect vectors. Entry is through wounds and natural openings. Initially, spots are black and round and surrounded by bacterial ooze. Under favourable conditions, spots enlarge to become raised, dark brown lesions with indefinite margins that cause the fruit to crack. The disease may cause up to 90% yield reduction. Severity increases during June and July and reaches a maximum in September and October and then declines. Bacterial cells are capable of surviving in soil for >120 days and also survive in fallen leaves during the off-season. High temperatures and low humidity or both favour disease development. Optimal temperature for growth of bacterium is 30°C; thermal death point is about 52°C.

Management

Cultural control

- New orchards should be established with certified disease free planting material or tissue culture raised saplings.
- Clean cultivation and strict sanitation in orchard.
- Pruning and burning of affected branches should be done regularly.
- Follow orchard sanitation measures strictly. Fallen twigs, leaves and Fruits should be destroyed outside the orchard premises.
- In bacterial blight prone areas only *hasta bahar* or late *hasta bahar* crop is recommended.

Chemical control

- Spray Bordeaux mixture (1.0%) during dormancy.
- Spray with Streptocycline (0.025%) in combination with Copper Oxychloride (0.25%) or Carbendazim (0.15%) at 15 days interval for 5-6 times starting from leaf initiation stage during rainy season and post-rainy season.
- If possible, cut ends should be pasted with Bordeaux (10%) paste after every pruning.
- Copper formulations + Streptocycline or Carbendazim + Streptocycline 0.05% and other bactericides if disease pressure is high and weather conditions are favourable.

Emergency Measures for Bacterial Blight Management

In case of sudden increase in BB on fruits take 3-4 sprays at 5 days interval

Spray 1: Copper hydroxide (2.0g/l) + Streptocycline (0.5g/l) + 2-bromo, 2-nitro propane-1, 3-diol (Bronopol) @ 0.5g / l + spreader and sticker (0.5ml/l)

Spray 2: Carbendazim (1g/l) + Streptocycline (0.5g/l) + 2-bromo, 2-nitro propane-1, 3-diol (Bronopol) @ 0.5g/ l + spreader and sticker (0.5ml/l)

Spray 3: Copper oxychloride (2.0g/ l) + Streptocycline (0.5g/l) + 2-bromo, 2-nitro propane-1, 3-diol (Bronopol) @ 0.5g / l + spreader and sticker (0.5ml/l)

Spray 4: Mancozeb 2g/l) + Streptocycline (0.5g/l) + 2-bromo, 2-nitro propane-1, 3-diol (Bronopol) @ 0.5g / l + spreader and sticker (0.5ml/l)

2. Anthracnose (*Colletotrichum gloeosporioide* Penz.)

Anthracnose is an important disease which affects leaves and fruits. Small, regular to irregular black spots on leaves, calyx region and fruits which turn later on as dark brown depressed spots. Infected leaves turn yellow and drop off. The disease is severe during August-September when there is high humidity, and the temperature between 20-27°C.

Fig. 3.13 : Symptoms of anthracnose on fruit

Management

Cultural control

- Select Haste or Ambe bahar varieties.
- Wider plant spacing, yearly pruning of trees.
- Proper disposal of diseased leaves, twigs and fruits.

Chemical control

- Spray the crop with Carbendazim (0.15%) or Mancozeb (0.25%) or Copper Oxychloride (0.25%) before plucking fruits at flower initiation during May to December.
- Apply Kitazin 48% EC @ 0.20% or 80ml in 80 l of water as required depending upon crop stage and plant protection equipment used.

3. Cercospora Fruit Spot (*Cercospora* sp.)

The affected fruits show small irregular black spots, which later coalesce into large spots measuring 1–12 mm diam. Leaf spots/lesions, are sub circular to irregular, 1–4 mm diam., at first brown, dingy gray to pale tan and eventually brown to dark brown at the margin.

Fig. 3.14 : *Cercospora punicae*
(See colour version on page 179)

Management

- The diseased fruits should be collected and destroyed.
- Two to three spraying at 15 days interval of Mancozeb (0.25%) after fruit formation gives good control of the disease.

4. Fruit Rot (*Phytophthora* spp., *Aspergillus foetidus* Thom & Raper)

The symptoms are in the form of round black spots on the fruit and petiole. The disease starts from calyx end and gradually the entire fruit shows black spots. The fruit further rots emitting a foul odour.

Management

- Spray the crop with Carbendazim (0.15%) or Mancozeb (0.25%) or Copper Oxychloride (0.25%) before plucking fruits at an interval of 10-15 days from the onset of flowering.

5. Alternaria Fruit Spot (*Alternaria-alternata*)

Small reddish brown circular spots appear on the fruits. As the disease advances these spots, coalesce to form larger patches and the fruits start rotting. The arils get affected which become pale and become unfit for consumption.

Fig. 3.15 : *Alternaria* fruit

Management

- All the affected fruits should be collected and destroyed.
- Spraying Mancozeb (0.25%) effectively controls the disease.

6. Wilt Complex (Complex of fungal infections *Ceratocystis fimbriata, Fusarium oxysporum*)

Pomegranate wilt complex is an important disease which results in complete wilting of plant. Affected plants show yellowing of leaves in some twigs or branches, followed by drooping and drying of leaves. The entire tree dies in few months or a year when affected tree is cut open lengthwise or cross-section dark grayish-brown discolouration of wood is seen.

Fig. 3.16: Wilt affected tree

(See colour version on page 179)

Causes

Ceratocystis fimbriata, Meloidogyne incognita, shot hole borer –*Xyleborus spp.* major cause and species of *Fusarium, Rhizoctonia, Sclerotium, Macrophomina, Phytophthora* are occasionally associated.

Predisposing factors

The wilt diseases are generally aggravated due to biotic stresses particularly drought as well as excessive rain, boron deficiency in soil result in increased severity to *C. fimbriata*. Wounds natural or due to insect/nematode or human activity like pruning and inter cultural operations predispose the plants to severe infections, as the pathogen is more devastating in overindulged orchards rather than in orchards with little human activity. Stress due to flowering and fruit bearing trees also results in sudden death of the entire plant. All commercial cultivars are susceptible to wilt and it can attack plants of all ages.

Management

1. The planting material (sapling as well as soil in which it is planted) should be free from all wilt causing agents-the fungi, insects and nematodes; use solarized/sterilized soil for planting saplings. It is advisable to take cuttings/ air layers from disease free orchards and make your own saplings using sterilized soil.

2. The soil used for potting mixtures or soil of beds for planting new orchards should be sterilized using chemical sterilants @ 2.5-5% formalin or 6 weeks of soil solarization using 50-100μ thick linear low density polyethylene (LLDPE) sheet during hot summer months. If formalin is used ensure that the soil is free from any formalin fumes before transplanting in the bags. Soil solarization is beneficial as it kills harmful pests and pathogens and also increases population of beneficial microrganisms which are present in the soil and are thermo tolerant, whereas, formalin treatment kills both harmful and beneficial organisms.

3. The above formalin treatment can be also be used for sterlizing soil after removing dead plant.

4. On observing first symptoms of wilt first ascertain the cause/s. If it due to fungal pathogens in the orchard immediately drench soil with Propiconazole 25EC (2ml/l) + Chlorpyriphos 20EC (2.ml/l) or Carbendazim 50WP (2.0g/ l) + Chlorpyriphos 20EC (2.ml/l) use 5-10 l solution/plant depending on growth so that 12 inches depth below shaded area becomes wet.

5. Also drench at least 3-4 healthy plants on all the four sides around the infected plant/s, repeat the drenching 3-4 times at 20-25 days interval. Drenching with Ridomil, metalaxyl or dithane M-45 (2g/l) will be beneficial if *Phytophthora* is causing any loss.

6. For controlling shot hole borer (*Xyleborus* spp.) which is associated with wilt disease, 10 litres preparation containing red soil (4 kg) + Chlorpyriphos 20 EC (20 ml) + Copper Oxychloride (25 g) needs to be applied on plant base up to 2 ft. from second year onwards. To control stem borer, inject in the holes on the trunk with DDVP 2-3 ml and plug the holes with mud.

7. Wilt due to **root knot nematodes** can be managed with soil application of Carbofuran 3G @ 20-40g/plant or other suitable nematicide in the plant basin, in a ring near root zone and cover it with soil. Drenching with azadirachtin (1%) @ 2ml/l is also recommended. Application of neem cake 1-3kg/plant depending on age is advisable twice a year. Plant *Tagetes erecta* (African marigold varieties best followed by French marigold) between plant to plant space in a row, or in a ring, on the border of plant basin. For effective results these should be grown for more than 4-5 month. Crops like onion, tomato, chilli, potato, capsicum, gram, legumes, cucurbits, Gerbera, Gladiolus *etc.* aggravate nematode infestation and hence should be avoided as intercrop. Green manuring with *sesbania* is beneficial.

Biological Control

1. Biological formulations if used should be reliable, fresh and used during rest period when no other fungicides/bactericides are used. The soil application of *Bacillus subtilis, Paecilomyces lilacinus, Pseudomonas fluorescens*, *Trichoderma harzianum*, *Aspergillus niger* 10-15g/plant along with well-decomposed farm yard manure around the trunk of pomegranate trees helps to prevent wilt infections. Neem cake @ 2-3 kg/plant effectively checks incidence of wilt complex.
2. Biofertilizer – Kalisena SA having *Aspergillus niger* @1 kg/acre+ Mychorrhizal preparation Josh @ 5kg/acre or Josh ultra 1Kg/acre- gives effective control of wilt if use from beginning or before disease starts. These two biofertilizers should be applied twice a year along with sufficient organics for effective wilt management. These controls several soil pathogens and also improves nutrient uptake and gives disease resistance and improves yields.

General precautions

1. Once disease is detected in the orchard, dig about 3-4 feet long trench between the wilted and healthy plant/s. The partially wilt affected plant/s should be treated with a suitable soil application depending on pathogen involved.
2. Dead plants should be removed and burnt, they should not be kept dumped in the orchard for firewood. While removing the wilted plants from the orchard for burning, protect the entire root zone with cover-fertilizer bag *etc.*, so that pathogens in soil on root do not spread in orchard.

3. The soil in the pit from which dead plant has been removed, should be sterilized with 2.5-5% formalin using about 10 l solution. It should be covered with polyethylene sheet for 1 week. After 1 week remove polyethylene sheet and rake the soil daily up to 10-15 days, so as to allow escape of gas. Plant new sapling once there is no smell of formalin in soil.
4. Pruning tools should be disinfected and cut ends painted with fungicidal oil based paints. Pruning should be avoided during spring to summer and done in winter months. Partially affected plants within the buffer zone should be treated with a suitable treatment; neighboring asymptomatic apparently healthy plants should also be treated with appropriate systemic fungicide/insecticide. Plants with more than 30% canopy loss should not be treated, they should be uprooted and burnt, soil treated with formalin and new plant grown.

3.3 Physiological Disorders

Fruit splitting and sunburn may affect pomegranate fruits and some-times cause significant commercial damage. Fruit splitting actually may be regarded as the last stage of normal pomegranate fruit developmental process where the fruit is spreading its seeds. Most known pomegranate cultivars will eventually split when they over-ripen. Some cultivars, such as the Israeli cultivars 'P.G.131-32', and 'P.G.118-19' and some cultivars from the pomegranate collection in Saveh (Tabatabaei and Sarkhosh 2006), tend to split in much earlier stages of fruit development or in higher frequencies than others. 'Shirvan', 'Burachni', and 'Asperonskii Krasnyi' were found to be resistant to splitting (Trapaidze and Abuladze 1998), suggesting that at least some aspects of fruit splitting in pomegranates are genetically controlled independently from environmental conditions. This assumption is also corroborated by Hepaksoy *et al.* (2000). The extent of fruit splitting is significantly reduced by regular irrigation, particularly by drip irrigation (Prasad *et al.* 2003). It is known that rainfall on mature pomegranate fruits following the end of the dry season can induce rapid fruit splitting. Therefore, splitting is a problem in regions where the fruit maturation overlaps a rainy season. There are indications from Israel and from Turkey that shading may induce fruit splitting,

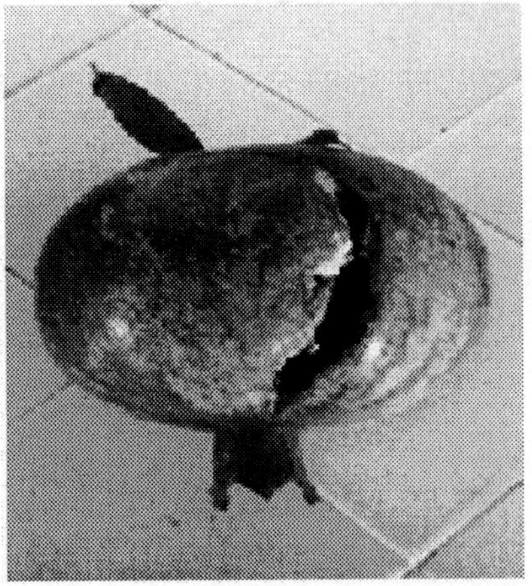

Fig. 3.17 : A view of a fruit cracking

most probably by changing the water balance due to lower radiation (Yazici *et al.* 2006). A few reports indicate that spraying with gibberelic acid (GA_3) at 150 ppm or with benzyl adenine (BA) at 40 ppm could significantly reduce fruit splitting (Sepahi 1986; Mohamed 2004). Other studies indicate that application of boron may reduce fruit split (Singh *et al.*, 2003; Shiekh and Rao 2006). In Israel, late cultivars such as 'Wonderful' that ripen in autumn and are exposed throughout the summer to strong solar radiation and hot temperatures are much more susceptible to sunburn. Early cultivars such as 'Akko' and 'Shani-Yonay' are less susceptible. It is not yet known whether there are differences among cultivars with respect to sunburn sensitivity and whether skin colour is a factor in this respect. Studies conducted by Yazici et al. indicated that 35% shading and application of Kaoline are effective in reducing sunburn on pomegranate fruit (Melgarejo *et al.*, 2004).

Management

1. The water retention capacity of the plants should be increased by the use of organic manures.
2. The plants should be irrigated regularly during the entire fruit development stage.
3. Spraying of Boron (50 ppm) and GA (40 ppm) on the young fruits minimizes the incidence of fruit cracking.

3.4 Integrated Disease and Insect Pest Management (IDIPM)

Schedule for Pomegranate Cultivation

This schedule can be used in general for getting disease and insect free yield in any season, however, farmers should not use Streptocycline (streptomycin sulphate 90% + oxy tetracycline 10%) and Bronopol (2-bromo, 2-nitropropane-1, 3-diol) if their orchards are free from bacterial blight and are in bacterial blight free areas. The spray interval of 7 days should be adopted in *mrig bahar* and 10-14 days in *hasta* and *ambe bahar* season.

Table 3.1 : Spray schedule for pest management

S. No.	Days after defoliation	Stage	Pesticides
Sprays During Crop Period			
1.	0-7	Defoliation	• Spray 1% Bordeaux mixture before defoliation
2.	8-14	85-100 % leaf fall	• Remove fallen leaves and debris from the orchard and burn • Drench soil with bleaching powder (33% Cl) @25Kg/1000 liters/1 ha • Spray Copper Oxychloride 50WP (2.5g/l) + Bronopol (0.5g/l)
3.	15-21	First flush of leaves	• Morning spray salicylic acid formulation @a.i.0.3g/l Evening spray Streptocycline (0.5g/l) +Carbendazim 50WP (1g/l) + Thiamethoxam 25WG @ 0.3g/l
4.	22-28	Flower initiation	• Bronopol (0.5g/l) + Ziram 80% WP 2g/l
5.	29-35	Flowering	• Morning spray mixture of zinc sulphate (3g/l), Solubor (2g/l), chelated iron (3g/l) • Evening spray Streptocycline (0.5g/l) + Carbendazim 50WP (1g/l) + Acetamiprid 20SP @ 0.3g/l/
6.	36-42	Flowering	• Spray Captan 50WP(2.5g/l) + Bronopol (0.5g/l)
7.	43-49	Flowering 100%	• Spray Streptocycline (0.5g/l) + Mancozeb 75% WP (2g/l)+ Imidacloprid 17.8SL @ 0.3ml/l
8.	50-56	Fruit set starts	• Morning spray salicylic acid formulation @ a.i. 0.3g/l Evening spray Streptocycline 0.5g/l) + Propineb 70WP (3g/l) or Ziram 80% WP(2g/l)
9.	57-63	Fruit setting	• Morning spraySolubor 2 g/lit + commercial micronutrient mixture 1g/l • Evening spray Steptocycline (0.5g/l)+Thiophanate Methyl 70WP (1g/l) + Cypermethrin 25%EC (1 ml/l) + Neem Seed Kernel Extract @50g/l (75g if entire seed is used) in evening
10.	64-70	Fruit setting	• Morning spray Magnesium sulphate (2g/l)Evening spray Bronopol (0.5g/l) +Fosetyl Al 80% WP (2g/l)
11.	71-77	Fruit enlargement	• Morning spray salicylic acid formulation @ a.i.0.3g/l Evening spray Bordeaux mixture (0.5%)

12.	78-84	Fruit enlargement	• Morning spray calcium nitrate (15 g/l)Evening spray Bronopol (0.5g/l) + Mancozeb 75% WP (2g/l)
13.	85-91	Fruit enlargement	• Spray Streptocycline (0.5g/l) + Carbendazim 50WP (1g/l) + methomyl 40%SP@ 1g/l +Neem seed kernel extract @50g/l (75g if entire seed is used)
14.	92-98	Fruit enlargement	• Morning spray salicylic acid formulation @ a.i.0.3g/l Evening spray Bronopol (0.5g/l) (0.5g/l) + Ziram 80% Wp (2g/l) + Azadirachtin 10000 ppm (3ml/l)
15.	99-105	Fruit enlargement	• Spray Bordeaux mixture (0.5%)
16.	106-112	Fruit enlargement	• Spray Steptocycline (0.5g/l) (0.5g/l) + Mancozeb 75% WP (2g/l)
17.	113-119	Fruit enlargement	• Spray Captan 50WP(2.5g/l) + Bronopol (0.5g/l)+ Methomyl 40%SP@ 1g/l· Drench with bleaching powder (33% Cl) @25Kg/1000 liters/1 ha
18.	120-126	Fruit enlargement	• Morning spray Calcium nitrate (8-10 g/l) Evening spray Steptocycline (0.5g/l) (0.5g/l) + copper hydroxide 77WP (2g/l) 77% WP (2g/l)in evening same day
19.	127-133	Fruit enlargement+Aril colour development	• Spray Bronopol (0.5g/l) + Thiophanate Methyl 70WP (1g/l) +Acetamiprid 20SP@ 0.3g/l+
20.	134-140	Fruit enlargement+Aril colour development	• Spray Steptocycline (0.5g/l) (0.5g/l) + Propineb 70WP (3g/l)
21.	141-147	Fruit enlargement and development	• Spray Bordeaux mixture (0.5%)
22.	148-154	Fruit enlargement and development	• Spray Bronopol (0.5g/l) + Captan 50% WP (2.5g/l)
23.	155-161	Fruit enlargement and development	• Spray Steptocycline (0.5g/l) (0.5g/l) + Copper Oxychloride 50WP (2.5g/l)+Lambda cyhalothrin 5EC/CS (0.5g/l)
24.	162-168	Fruit enlargement and development	• Bordeaux mixture (0.5%)
25.	169-184	Fruit enlargement & development	• Spray Potassium dihydrogen phosphate @ 10g/l + Spray Bronopol (0.5g/l) + Captan 50% WP (2.5g/l)
26.	185-199	Fruit Maturity	• Spray Neem seed kernel extract @50g/l (75g if entire seed is used) or Azadirachtin 1500ppm @ 3ml/l
27.	200-214	Fruit Maturity (1 month before harvest)	• Spray Potassium nitrate@10g/l or 0:0:50 @10g/l Bordeaux mixture @ 0.5% only under adverse weather conditions
28.	215-230	Fruit ripening	• Harvest

Sprays During Rest Period

- Apply Bordeaux paste (10%) on pruned ends. Immediately after pruning spray Bordeaux Mixture (1%).
- Spray Bordeaux Mixture (1%) at 15 days interval or alternate with Bronopol @0.5g/l+ Copper Oxychloride 50WP (2.5g/l) spray. Continue through defoliation.

Points to remember

- A pomegranate surfaces are glossy, hence, preferably add good quality non non-ionic spreader sticker with sprays for uniform coverage with pesticide. Do not use spreader sticker with bordeaux mixture.
- To prepare spray mixture, prepare dilute solutions of each chemical separately and mix to make total volume. If precipitate is formed, either mixture chemicals are not compatible or pH is not proper.
- The spray solution should have a pH of 6.5 -7 for good results.
- In case no rains are there for long duration or blight is not increasing, sprays can be taken at 10-15 days interval instead of 7 days.
- The active ingredients (a.i.) in Streptocycline are streptomycin sulphate 90%+oxy tetracycline 10% and in Bronopol '2-bromo, 2-nitropropane-1, 3-diol 95%.' Both Streptocycline and bronopol are available with different trade names from different companies. check if a.i. is less then increase the dose accordingly.

Precautions

- Take only need based sprays at recommended doses, too many sprays increase the disease.
- Always remove and burn all affected fruits before starting any spray.
- Combine insecticides, fungicides or micronutrient sprays with bactericidal sprays depending on compatibility to reduce number of sprays. Mixture should not form precipitate.
- Take without fail, additional spray with a bactericide after the rains -when plant surfaces dry up.
- Always prepare Bordeaux mixture fresh and use on the same day

Table 3.2 Cultural operations

S. No.	Days after defoliation	Stage	Cultural Operation
1.	0-7	Defoliation	• Defoliate with Ethrel (1.5-2ml/l)+DAP5g/l • Remove weeds and suckers • Apply 2/3rd dose of FYM + Micronutrients depending on deficiency + poultry manure (1-1.5 kg/plant), vermicompost (1Kg/plant), Neem cake (1.5-3Kg/plant)Neem Cake 1-2 Kg /plant + Vermicompost 1 Kg/plant + Phorate10G @25g/plant or Carbofuran 3G @ 40g/plant in wet soil in a ring around the plant • Apply 1/4th dose of nitrogen • Give light irrigation immediately after fertilizer application
2.	8-14	85-100 % leaf fall	• Remove fallen leaves and debris from the orchard and burn • Drench soil with bleaching powder (33% Cl) @ 25Kg/1000 liters/1 ha
3.	15-21	First flush of leaves	• Irrigate
4.	22-28	Flower initiation	• Apply 1/4th dose of nitrogen
5.	29-49	100% Flowering	• Fertigate N:P:K::12:61:00 @ 8 kg/ha/ application. Give 15 applications on alternate days for a month through irrigation. • Irrigate
6.	50-63	Fruit set starts	• Apply 1/4th dose of nitrogen in soil • Fertigate N:P:K::19:19:19 @ 8 kg/ha/ application.
7.	64-70	Fruit setting	• Give 15 applications on alternate days for a month through irrigation • Remove weeds Irrigate regularly
8.	71-126	Fruit set 100%Fruit enlargement	• Fertigate N:P:K::00:52:34 Mono-Potassium Phosphate @ 2.5 kg/ha/application • Give 15 applications on alternate days for a month through irrigation • Apply 1/4th dose of nitrogen in soil in August first week • Irrigate • Drench soil with bleaching powder (33% Cl) @25Kg/1000 liters/1 ha in last week of Aug. or first week of Sep.

9.	127-140	Fruit enlargement+Aril colour development	• Remove weeds and suckers • Irrigate
10.	141-184	Fruit enlargement & development	• Remove weeds and suckers • Irrigate
11.	185-199	Fruit Maturity	• Remove weeds and suckers • Irrigate
12.	200-214	Fruit Maturity 1 month before harvest	• Fertigate Calcium Nitrate @12.5 kg/ha application give 2 applications at 15 days interval • Give moderate Irrigation
13.	215-230	Fruit ripening	• Harvest mature fruits
Operations During Rest Period			
14.	-	Rest	• Do heavy pruning to remove blight affected and old dry branches • Drench soil with bleaching powder (33% Cl) @25Kg/1000 liters/1 ha • Apply 1/3rd dose of FYM + 1/4th dose of nitrogen + Neem cake @ 1Kg /plant and full dose each of phosphorus and potash, poultry manure (1-1.5 kg/plant), vermicompost (1Kg/ plant), Neem cake (1.5- 3Kg/plant) • Light Irrigation
15.	-	Rest	• Light Irrigation
16.	-	Stress	• Stop Irrigation

Note

- Irrigate depending on soil type, plant age and stage and weather conditions.
- Apply nitrogen, phosphorus and potash depending on plant age.

Source : http://www.nrcpomegranate.org/documents.aspx

4

Harvesting and Post Harvest Management

4.1 Harvesting

Pomegranate being non-climacteric fruits should be picked when fully ripe. Its fruits become ready for picking 120-130 days after fruit set. Mature fruits should be immediately removed from the plants as delay in harvesting leads to fruit cracking. Fruits harvested at a premature stage have poor keeping quality and are prone to damage during handling and transport. The fruits are harvested when outer rind becomes yellowish and the fruit when tapped produces a metallic sound. The round fruits become flattened from all the sides and basal beak shaped portion shrinks at the time of maturity. Fruits are harvested with the help of secateurs by retaining 1 cm stalk with the fruit. All the fruits should be harvested in 2-3 pickings within a span of one month. The harvested fruits are cured for a week in shade. This will make the skin harder and will stand better in transportation.

Fig. 4.1 : Fruits ready for harvest

(See colour version on page 180)

4.2 Harvesting Season of Crop in Leading States

Harvest season of pomegranate is depicted below (in 12 months). Because of adopting a number of bahar treatments, pomegranate in Maharashtra and Gujarat states is available throughout the year.

	Lean Period				Peak Period					Throughout Year		
State/UT'S	**Jan**	**Feb**	**Mar**	**Apr**	**May**	**Jun**	**Jul**	**Aug**	**Sep**	**Oct**	**Nov**	**Dec**
Maharashtra												
Karnataka												
Andhra Pradesh												
Gujarat												

*The above graph showing harvest pattern in leading Pomegranate growing states.

Yield

The plant starts bearing from the 3rd year onwards. A grown-up, well-managed tree gives 60-80 (25-30 Kg) fruits annually, with a life span of 25-30 years.

4.3 Arrival Pattern in Market

Pomegranates are available almost throughout the year. With the adoption of bahar treatment it's harvest can be tailored according to demand.

Table 4.1: Details of arrival pattern of pomegranate according to bahar treatment

S. No.	Bahar	Flowering Time	Period of Harvest
1.	Mrig	June-August	November-March
2.	Hasta	October-November	February-May
3.	Ambe	January-February	June-August

Catchment Areas of Market

Table 4.2 : Showing the details of catchment areas of markets of pomegranate in leading states

States	Districts (Market)	Blocks
Maharashtra	Solapur	Karmala, Barsi, Madha, Mohol, Mangalwedha, Śingole, Malsira, Pandharpur, Akalkot
	Nasik	Kalvan, Peint, Igatpuri, Sinnar, Niphad, Yeola, Nandgaon, Satana Furgana, Dindori, Melgaon
	Sangli	Atpadi, Khanapur, Islampur, Shirala, Valva, Tasgaon, Kavathe, Mahankal, Jath, Miraj
	Ahmednagar	Srirampur, Sangamner, Akola, Rahuri, Nevasa, Parner, Pathardi, Srigonda
	Pune	Junnar, Ambegaon, Ghod, Rajgurunagar, Wadgaon, Sirur, Mulshi, Welhe, Purandhar, Bhor, Baramati, Indapur, Daund, Saswad
	Satara	Mahabaleshwar, Khandala, Wai, Phaltan, Koregaon, Khata, Patan, Karad, Vadug
Karnataka	Bijapur	Indi, Sindgi, Basavna Bagevadi, Muddebihal, Tikota
	Belgaum	Athni, Arkali, Chikodi, Mukeri, Bailhongal, Ramdurg
	Bagalkot	Jamkhandi, Mudhol, Hungund, Badami
Andhra Pradesh	Anantpur	Guntakal, Gooty, Rayalacheruve, Uravakonda, Kanekallu, Rayadurg, Kalyandurg, Kambadur, Manakasira, Nallamada
Gujarat	Bhavnagar	Botad, Gadhda, Valbhipur, Umrala, Mandir, Gariadhar, Palitana, Talaja, Chogha, Vaibhipur
	Ahmedabad	Mandal, Rampura, Samand, Bavia, Dholka, Dhandhkulla, Ranapur
	Sabar- Kantha	Vijarnagar, Khedbrahma, Vadali, Idar, Bhiloda, Talod, Dhansura, Bayad, Malpur, Meghraj, Bhiloda

4.4 Post Harvest Management

A review on postharvest biology and technology of pomegranate was recently published by Kader (2006). The review summarizes the current knowledge on the pomegranate morphological characteristics, composition and compositional changes during maturation and ripening, quality indices, and postharvest biology. This section will mention some of the major points concerning postharvest biology of the pomegranate fruit in addition to the new postharvest technologies developed and practiced today in Israel.

As the pomegranate fruit matures on the tree, a reduction in the titratable acidity and parallel increase in TSS, pH, and colour intensity is observed (Kader 2006). Once the fruit is harvested, it keeps respiring at a relatively low rate. This rate is decreased with time after the harvest. Storage at low temperature can keep respiration rate under 8 ml CO_2 per kg min^{-1}. Due to relatively low respiration rates and low amount of ethylene evolved (0.2 ml per kg min^{-1}), the pomegranate fruit is classified as nonclimacteric. The pomegranate fruit increases its respiration rate and ethylene production immediately after exposure to ethylene. However, the effect of ethylene treatment on respiration rapidly declines. Ethylene treatments did not affect significantly fruit parameters of harvested fruit such as colour, TSS, pH or titratable acidity. These data indicate that the pomegranate fruit will not mature postharvest and should be harvested only when fully mature.

The major problems in pomegranate storage are loss of fruit weight, fruit size reduction, skin damages such as husk scald (browning of the skin surface) (Or-Mizrahi and Ben-Arie 1984; Ben-Arie and Or 1986; Defilippi *et al.,* 2006), and development of crown and fruit rot (Adaskaveg and Forster 2003; Tedford *et al.,* 2005).

Gray mold caused by *Botrytis cinerea* Whetzel and rot caused by *Penicillium implicatum* Biourge 1923, Rhizopus arrhizus Fischer 1892, and *Alternaria solani* Sorauer 1896 are storage diseases of pomegranate fruit (Kanwar and Thakur 1973; Vyas and Panwar 1976; Morton 1987; Labuda *et al*., 2004; Palou *et al*., 2007). In California, Botrytis cinerea, which causes postharvest decay, is the primary limiting factor for long-term storage (Adaskaveg and Forster 2003; Tedford *et al*., 2005). Fenhexamid and fluidioxonil treatments were shown to be very effective in reducing natural incidence of gray mold caused by *B. cinerea.* To prevent development of fungicide resistance in these pests, a combination of sanitation treatments with chlorine and fungicides dip was recommended before cold storage (Adaskaveg and Forster, 2003). Palou *et al.* (2007) indicated synergistic effects between antifungal treatments and controlled atmosphere (CA) of 5 kPa O_2 þ 15 kPa CO_2 in 'Wonderful' pomegranates artificially inoculated with B. cinerea. A combination of waxing with antifungal treatments was suggested by Ghatge *et al.* (2005) to extend the shelf life and the quality of pomegranate in cold storage and ambient conditions.

Pretreatment of pomegranates with hot water at 45°C was shown to reduce chilling injury and electrolyte and K leakage (Artes *et al*., 2000; Mirdehghan and Rahemi 2005). Heat treatment was also shown to be effective in maintaining the nutritive and functional properties of pomegranate fruit after a long period of storage (Mirdegahan *et al.,* 2005) and in reducing pomegranate moth damage (Moghadam and Nikkhah, 2005).

Pomegranate fruits can be kept well for long time after harvesting. Experimental data showed that fruits could be stored betwe2en 0°C and 4.5°C at 85% relative humidity for several months (Mukerjee 1958; Kader *et al.*, 1984; Or-Mizrahi and Ben-Arie 1984). Saxena *et al.* (1987) found that time of harvest, temperature, and oxygen level significantly affected husk scald. These authors found that delaying harvest reduced the percentage of fruit developing husk scald. Low oxygen and low temperatures inhibited husk scald, probably by inhibiting enzyme-dependent oxidation processes (Or-Mizrahi and Ben-Arie 1984). A combination of low oxygen levels (2%-3%), low temperature (2-6°C), and late harvest were found optimal to reduce chilling injuries while preserving taste qualities. Low oxygen levels may cause anaerobic respiration, which in turn causes the accumulation of fermentive volatiles (Kader *et al.* 1984; Or-Mizrahi and Ben-Arie 1984). Therefore, 5% oxygen was suggested as a compromise oxygen level where skin damage is inhibited while fermentive volatiles are not produced (Kader, 2006). The effectiveness of CO_2 in inhibiting scald development in addition to its fungicidic effects led Hess-Pierce and Kader (2003) to recommend 5% oxygen þ 15% CO_2 as the optimal CA for pomegranate storage at 7°C and 90% to 95% relative humidity (Kader, 2006). Ranjbar *et al.* (2006) demonstrated that polyethylene bag wraps significantly reduced weight loss and improved appearance of the fruit following storage. In Israel, several long storage experiments were conducted by Porat *et al.* (2005, 2007). These authors have developed new storage technology (modified atmosphere packaging) that involves the usage of special bags (Xtend[1]) which have small pores (microperforation) (Sachs *et al.*, 2006). These bags result in the development of 5% CO_2 and 12% to 14% O_2 within the bag surrounding the fruit. The Xtend packaging reduces weight loss from 7% to 3.5%, reduces scald from 38% to between 2% and 11%, and reduces crown decay when pomegranate fruit were stored at 6°C for 16 weeks. Using either the Xtend packaging technique just described or CA conditions of 2% O_2 þ 3% CO_2 at 6°C permitted storage of pomegranate fruit for to four to five months with acceptable commercial quality. Antifungal pretreatment of the pomegranate fruit was recommended before storage began. Data on storage experiments that was reported by Hess priece and Kader (2003) and Porat *et al.* (2005) were based on the cultivar 'Wonderful'. As mentioned in the cultivar section, the fruit qualities of the Israeli 'Wonderful' and the American 'Wonderful' are different. It was also demonstrated that different pomegranate cultivars contain different levels of secondary metabolites that have antioxidant activities (Tzulker *et al.*, 2007). These in turn could potentially change the sensitivity of pomegranate fruit to skin damage and pathogen attack. Therefore, care should be taken with respect to storage conditions in each geographical region and for each cultivar. Currently many new cultivars are being introduced to commercial growth in addition to 'Wonderful'. There-fore, special postharvest experiments should be carried separately for each cultivar.

Apart from consumption of pomegranates as fresh fruit, they are also used for other purposes, such as isolated arils, juice, wine, and health-promoting agents. One of the newest developments in pomegranate culture is an efficient commercial method to extract intact arils (Rodov *et al.*, 2005; Shmilovich *et al.*, 2006). Several machines were developed, but the most efficient one is able to produce more than 1 ton of arils a day (Shmilovich *et al.*, 2006). This development requires some new studies in order to prolong the shelf life of the arils and to preserve them either as fresh or frozen product. Such studies are only just beginning. Juice is produced industrially from either crushing whole pomegranate fruit or isolated arils. Some manufacturers preferred the isolated arils because the juice is less bitter and tastes better to many people. The by product of the aril and juice industry are the remnants of the fruit skin, membranes, and seeds. The fruit skins and membranes are rich in elagitannins, which have a wide array of health-promoting bioactivities (Seeram *et al.*, 2006) and their extracts have a commercial value for humans and for animal feed. The seeds are a source of oil that contains a rare combination of unsaturated fatty acids (Seeram *et al.*, 2006) and sterols. Seeds powder is a common component of some Indian food recipes.

4.5 Grade Designation and Quality

1. Minimum requirements

i. Pomegranates shall be:

   a. fresh in appearance;

   b. mature and solid in feel;

   c. clean, free from any visible foreign matter;

   d. free from pests affecting the general appearance of the produce;

   e. free of damage caused by pests;

   f. free of cracking of skin, mechanical injury/rubbing, staining;

   g. free of abnormal external moisture excluding condensation following removal from cold storage;

   h. free of any foreign smell or taste;

   i. free of any pronounced blemishes;

ii. Pomegranates should not be affected by rotting or deterioration such as to make it unfit for consumption.

iii. Pomegranates shall comply with the residue levels of heavy metals, pesticides and other food safety parameters as laid down by the Codex Alimentarius Commission for exports.

2. Criteria for grade designation

Table 4.3 : Details of grade designation and sizing of pomegranate as per AGMARK standard

Grade Designation	Grade Requirements	Grade Tolerances
Extra class	Pomegranate in this class must be of superior quality. They must have the shape, development and colouring that are typical of the variety and/or commercial type. They must be free of defects, with the exception of very slight superficial defects, provided these do not affect the general appearance of the produce, the quality, the keeping quality and presentation in the package.	5% by number or weight of pomegranates not satisfying the requirements of the grade, but meeting those of class I grade or, exceptionally, coming within the tolerances of that grade.
Class I	Pomegranates in this class must be of good quality. They must be characteristics of the variety and/or commercial type. The following slight defects may be allowed, provided these do not affect the general appearance of the produce, the quality, the keeping quality and presentation in the package.- a slight defect in shape.- a slight defect in colouring;- slight skin defects (i.e. scratches, scars, scraps and blemishes) not exceeding 5% of the total surface area	10% by number or weight of pomegranates not satisfying the requirements of the class, but meeting those of class II or, exceptionally, coming within the tolerances of that grade.
Class II	This class includes pomegranates which do not qualify for inclusion in higher classes, but satisfy the minimum requirements. Following defects may be there provided the pomegranates retain their essential characteristics as regard the quality, the keeping quality and presentation:- defects in shape;- defects in colouring- skin defects (i.e., scratches, scars, scrapes and blemishes), not exceeding 10% of the total surface area.	10% by number or weight of pomegranates not satisfying the requirements of the grade, but meeting the minimum requirements.

3. Other requirements

Pomegranates must be carefully picked and have reached an appropriate degree of development and ripeness in accordance with criteria proper to the variety and/or commercial type and to the area in which they are grown. The development and condition of the Pomegranate must be such as to enable them;

- to withstand transport and handling, and
- to arrive in satisfactory condition at the place of destination.

4. Provisions concerning sizing

Size is determined by the weight or maximum diameter of the equatorial section of the fruit, in accordance with the following table:

Table 4.4 : Details of sizing in pomegranate fruits

Size Code	Weight in grams (minimum)	Diameter in mm (minimum)
A	400	90
B	350	80
C	300	70
D	250	60
E	200	50

Size Tolerance

i. For all grades, 10% by number or weight of pomegranate corresponding to the size immediately above and/or below than indicated on the package.

ii. The maximum size range of 8 mm. between fruit in each package is permitted.

Source : Procedure for Export of Pomegranates, APEDA, New Delhi.

4.6 Packing

1. For export market

Usually for packing pomegranates for export purposes, a cardboard corrugated fibreboard box of 4.0 or 5.0 kg capacity is used. The details specifications are given below in the table:

Fig. 4.2 : Packing in carton box
(See colour version on page 180)

Table 4.5 : Specification details for corrugated fiber board (CFB) boxes for packing pomegranates for exports

Specification	Ring & Flap Tuck in type	RSC (regular slotted container)	Slide type
Material of Construction	5 Ply CFB	3 Ply CFB	5 Ply CFB
Grammage gm/m2 (outer to inner)	*230X140X 140X140X140	*230X140X 140X140X140	*230X140X 140X140X140
Bursting strength kg/cm2	Min 10	Min 10	Min 10
Puncture resistance, inches / tear inch	Min 250	Min 250	Min 250
Compression strength, kgf	Min 350	Min 350	Min 350
Cobb (30 min g/m2)	Max 130	Max 130	Max 130

* Outer ply of white duplex board
Source : Post- Harvest Manual for Export of Pomegranates, APEDA, New Delhi.

2. For domestic markets

For domestic markets also, the Pomegranate is packed in Corrugated Fiber Board boxes, according to their weight.

4.7 Storage

Fruits can be stored at 5°C with 90-95% relative humidity for 2 months. In case of storage beyond two months, temperature should be maintained at 10°C to avoid chilling injury. Pomegranates are very susceptible to water loss resulting in shrivelling of the skins. Storing fruit in plastic liners and waxing can reduce water loss, especially under conditions of lower relative humidity.

5

Exports and Export Procedure

5.1 Present Status

Export of pomegranate has decreased in quantity from 36,027.43 tons in 2012-13 to 20,997.02 tons in 2014-15, whereas in value term it show an increase trend during the same period. There is tremendous potential for exports of pomegranate from India and it is fact that India is largest producer of pomegranates in the world. Moreover, India produces finest edible quality of pomegranates which are available almost throughout the year. The major Markets of India's pomegranate during the year 2014-15 were UAE, Nepal, Saudi Arabia, Kuwait and Netherland.

Table 5.1 : Export of pomegranate from India in the last three years

Years	Quantity (tons)	Value (in Rs. Lakhs)
2012-13	36,027.43	23,449.62
2013-14	31,328.32	29,851.63
2014-15	20,997.02	32,361.43

Source : http://agriexchange.apeda.gov.in/indexp/Product_description.aspx?hscode=08109010

5.2 Export Potential

There are great prospects for increasing the pomegranate export to different countries. As compared to other countries where pomegranate supply is for a restricted period (fall and early winter), India can supply pomegranate to international market throughout the year due to multiple sessions of *bahar (Ambia, mrig & hasth bahar)*. In India, the peak production season is during December – March and it continues up to June-July. Thus, India can export pomegranate from February to July when there is no competition from other countries. To enhance export, increasing production of exportable quality fruits and providing postharvest handling facilities are required to be priority areas.

5.3 Export Destinations

Presently the major clients for the filed fresh are Middle Eastern countries and European countries. The specifications of the countries vary from one another the European exports need high quality produce compared to the Middle East and so the price offered by the company to the farmer for the two qualities. Field fresh should find new markets like Far East for its Pomegranates.

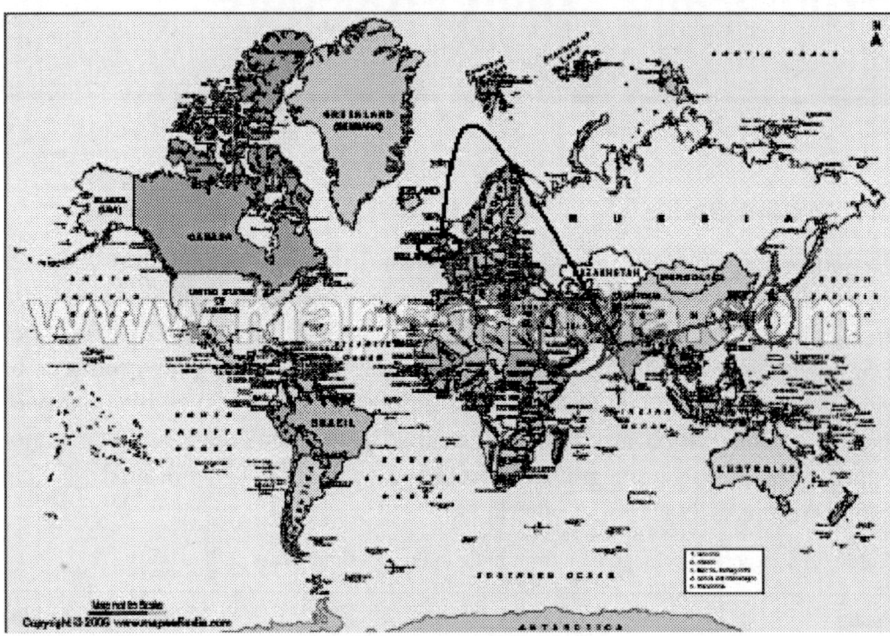

Fig. 5.1 Export destinations

5.4 Procedure for Export

Monitoring of residue levels of agrochemicals permitted for use by the Central Insecticide Board and Registration Committee (CIB&RC) for pomegranates is a major concern. It is essential that adequate monitoring through proper surveillance should be in place to eliminate the possibility of detection/presence of the residues of agrochemicals and any other contaminants in pomegranates in excess of prescribed levels of the importing countries. Accordingly, it is necessary to check/verify agro chemicals used in the cultivation of pomegranates for exports complying with food safety norms of importing countries. It is also essential to grade the pomegranates according to the AGMARK standards before issue of the Phyto Sanitary Certificate (PSC). In order to ensure implementation and compliance, following procedure would be followed:

1.	Objective	1.1	To establish a system for controlling residues of agrochemicals in exportable pomegranates at the farm and plot level.
		1.2	To monitor chemical residues in soil and water at pomegranate farms or plots and pack houses.
		1.3	To facilitate export of pomegranates by establishing a surveillance system for controlling residues of chemicals registered by CIB&RC and/or as recommended by the National Research Centre for Pomegranates (NRCP) during cultivation of pomegranates as well as for traces of other chemicals, which might be found due to previous use on the land.
		1.4	To establish a system for corrective action in the event of detection of residue levels higher than those established through these procedures as well as in the event of issuance of an Internal Alert Information.
		1.5	To ensure that pomegranates exported from India to the European Union (Austria, Belgium, Bulgaria, Cyprus, Czech Republic, Denmark, Estonia, Finland, France, Germany, Greece, Hungary, Ireland, Italy, Latvia, Lithuania, Luxembourg, Malta, Netherlands, Poland, Portugal, Romania, Slovakia, Slovenia, Spain, Sweden and United Kingdom) as well as other countries following EU food safety norms do not test for agro chemicals residues in excess of the prescribed levels.
		1.6	To establish a mechanism to provide grade classification of pomegranates through grant of Certificate of Agmark Grading (CAG) by the Department of Marketing and Inspection (DMI) with a view to ensuring export of quality pomegranates to the European Union Market as well as other countries following EU food safety norms.
2.	Scope	2.1	All pomegranates farms and plots, which intend to produce pomegranates for export, exporters, APEDA recognized pack houses, APEDA recognised laboratories authorized for sampling, analysis and grading of pomegranates and personnel engaged by them, NRL, NRC Pomegranates, Central Insecticide Board, DMI, respective State Government departments of Agriculture/ Horticulture, Indian Council of Agriculture Research (ICAR) and any other agencies/ stakeholders including extension providers in any form to the farmers, producers and exporters of pomegranates cultivating for exports.
		2.2	To facilitate web-based traceability through AnarNet with the objective of tracing and tracking of processes implemented product recall, single window clearance and reducing paper work.
3.	Procedure for registration of farms producing pomegranates for exports	3.1	Each farmer/lessee farmer/lessee farming organization who directly controls farming operations, who intends to export directly or supply pomegranates to an exporter, will apply for registration of its farm and plot(s) to the concerned District Superintending Agri./Hort. Officer, as per application form for registration of pomegranates farms given in Annexure-1.

3.2 After receiving the application from farmer/lessee farmer/lessee farming organization District Superintending Agriculture/ Horticulture Officer shall physically verify the correctness of the information submitted. Only then the same will be entered in the records maintained by the District Superintending Agri./Hort. Office for the purposes of this document, and not on the basis of any other records. He shall also verify that a plot(s) is/are not under suspension/cancelled for export. In the plot(s), the Agri./ Hort. Office shall take possession of the Registration Certificate from the farmer. The concerned officer shall be directly responsible for any incorrect/incomplete information entered in the records and in the registration certificate issued to the farmer.

3.3 Each farmer/lessee/lessee organization will maintain the registration and agrochemical application records as per the format titled, "Plot Registration and Field Agrochemical Application Record" given in the Annexure-2. The farmer/the lessee/the lessee organization and Agriculture/Horticulture officer must sign and write their complete official address. This is mandatory. The labs will not be able to test the samples for agrochemical residues if Annexure-2 is not completed by the farmers/lessee/lessee organization and a copy provided to the laboratory.

3.4 It shall be the responsibility of Agriculture/Horticulture officers to ensure that information required Annexure-2 is uniformly and completely maintained.

3.5 The Annexure-2 must describe the layout with detail of the adjoining farms/plots. A drawing/sketch of the layout, giving detailed description or benchmark including directions for identification, also showing details of late registration plots. Where applicable, should be enclosed with this Annexure (see paras 3.6, 3.13 and 4.3). The drawing/sketch must also be signed by the Agri/Hort Officer under his official sea.

3.6 The following guidelines are required to be followed by the Agri/ Hort Officer to prepare the drawing/layout of the plot(s) presented for registration/renewal :

a) indicate the directions (North, East etc.) of the plot,

b) indicate benchmarks, such as, roads, civil structures (hut, house, railway crossing, electrical sup-station, etc), canals, foot-paths, are of the plot where there are other plots (unregistered) or growing other crops), and

c) clearly indicate adjoining farms of other farmers.

3.7 A registration number will be given to each farm unto plot level, which will be under the charge of an Agriculture/Horticulture Officer whose head quarter is near the farm. The farm/plot would be allotted a registration number by adopting only the following code format:

State	District	Taluka/ Mandal	Product	Farm. No.	Plot No.
AA	11	22	33	4444	00

3.8 The plot registration certificate shall be issued only through Anar Net.

3.9 One plot will constitute a maximum area of Ha. Or part thereof. For marginal adjustments, the area may be extended to a maximum of 1.2 Ha Further, in case, different varieties are grown in the same plot or even if it is less than 1 Ha it will be treated as a separate plot and will have separate registration numbers.

3.10 If there is only one plot in a farm, even then the registration number will be unto the level by mentioning '00' at the end, e.g. AA-11 22 33 4444 00

3.11 Each plot in the farm shall be registered separately, e.g. AA-11 2233 4444 01 farm are not intended to be utilised for exports, it/ these will still have to be registered by the same procedure.

3.12 Amendments, if any, in the Registration Certificates, shall be made only by the authorised Govt. officials after due verifications and under their signature.

3.13 The registration will be valid for one year. Therefore, all the registered pomegranates farmers shall renew their farm registration every year and a fresh certificate showing renewal of the plot will be issued.

3.14 A farmer should apply for registration/renewal bahar wise at pruning time. Ambe Bahar: Dec-Jan, Mrig Bahar- May-June, Hasta Bahar- Aug-Sept.

3.15 A farmer/lessee farmer/lease farming organization should apply for renewal to the Registration Authority as per format given in Annexure-1. No application for registration of pomegranates farms/ plot(s) will be accepted after mentioned dates. However, if a farmer/lessee/lease organization expects a good crop and intends to export his produce, he should register the plot well before the maturity stage. The registration will be subject to the satisfaction of the Agri/Hort Officer with regard to maintenance of spray records and their uniformity with Annexure-2. In such cases, registration will be done after residue analysis of pomegranates (at farmer's cost/lessee farmer/lease farming organisation's cost) from such plot areas.

3.16 Registration Certificate, after complying with the procedure set out in paras 3.2 to 3.13, shall be issued to the applicant duly signed by an authorised State Government official indicating details of the plot, name of farmer/ lessee /lessee organisation, village, taluka, mandal, district, Survey No./Gat. No. (This is a critical information), variety wise area, age of plot(s) and package of practices to be followed by the farmer/lessee/lessee organisation for production of pomegranates.

a) The Registration/Renewal Certificate would be accompany by following instructions, which shall also be signed by the Agri/Hort Officer under his official seal to follow only the recommended farming practices.

b) That farmer/lessee farmer/lessee farming organisation shall not use agrochemicals other than those allowed for use on pomegranates and listed in Annexure – 5.

c) That farmer shall not use agrochemicals other than those allowed for use on pomegranates and listed in Annexure-5.

d) Misbranded, un-recommended or banned agrochemicals or any other harmful chemicals shall not be used.

e) After drawl of samples for residue testing, spraying/application of any agrochemicals shall not be carried out.

f) Not to allow sampling or exports of pomegranates from unregistered farms.

g) Amendments, if any, on the registration records or the Registration Certificate issuing authority.

3.17 All farmers/lessee farmer/lessee farming organisation shall maintain a record of package of practices followed by them in a prescribed register to be provided by the respective State Horticulture/ Agriculture Departments. This may include information on the cultural practices, application of fertilizer, dosage and date of application of insecticides, fungicides, weedicides & plant growth regulators and other chemicals.

3.18 Each farmer, at the time of harvest, shall give declaration to the exporter in Annexure-3 stating that pesticides, insecticides, fungicides, weedicides, etc. have not been sprayed/applied after drawl of samples for laboratory analysis. The declarations shall also state that there is no plot under the farm that is not registered by the Agri./Hort. Office and that none of the plots mentioned in the declaration are under suspension/have been cancelled for export. This declaration shall be handed over to the exporter at the time of harvesting of the produce.

3.19 The exporter shall, at the time AGMARK inspection, provide Annexure-4 to the laboratory representative.

4.	Procedures for District Superintendent Agriculture/ Horticul ture Officers	4.1	Each Agriculture/Horticulture Officer will visit the farm/plot(s) at least two times to inspect the farm/plot prior to harvesting of the pomegranates. The first inspection should be carried out at the time of farm/plot registration and the second one at the sampling. If the situation warrants due to potential or heavy pest/disease incidence, then an additional inspection should be made to such areas at a time considered appropriate by the Agri/Hort Officer. Each Agri/Hort Officer will prepare the report as per format of inspection report given in Annexure - 4 (A) & 4 (B) and give a copy to farmer/lessee farmer/lessee farming organisation after obtaining his signature on it. A copy of Annexure-4 (B) duly completed and signed by all concerned will be given to representative of the lab at the time of sampling.

4.2 In the interest of information flow in the AnarNet, the Annexure-4(B) is required to be filled up by the Agri/Hort Officer through the AnarNet, otherwise the farmers/lessee farmer/lessee farming organisation and exporters will face problems as the software will not move forward and lab analysis cannot take place.

4.3 In case of late registration of the plot(s) before ripening/maturity (please see para 3.13), the Agri/Hort Officer, during first inspection, will have to clearly demarcate such areas of the plots giving block-the area directions and/or locations (such as near the road, well etc.) This will also be indicated on Annexure-4 (A) and the Annexure-2 (and its enclosures) { Please see paras 3.5 & 3.6)

4.4 The Agri./Hort. Officer shall also verify the registration record of the entire farm and also physically check if the same is correct and complete in all respects.

4.5 The farmer/lessee farmer/lessee farming organization and the Agri./Hort. Officer must sign and write their complete official address on each form (Annexure 4-A to 4-B). This is mandatory.

4.6 The State Governments shall maintain plot registration data in the specified format in the web-based data-base on the Apeda website (www.apeda.gov.in) and prints of the registration certificates shall be taken through the web-based software placed on the website.

4.7 The Agri/Hort Officer shall not recommend drawl of samples by the laboratories in Column 15 of Form 4(B) if the farmer/lessee farmer/lessee farming organization has not followed the officer's advice/recommendation given in Form 4(A) or is, otherwise, satisfied that there is likelihood of the presence of excess pesticide residues. Reasons for not recommending drawl of samples will have to be clearly stated in column 16.

4.8 The recommendation for allowing drawl of samples shall be made by Agri/Hort Officer only through AnarNet. It is to be noted that, in the software, this recommendation can be made only after 30 days of registration of the plot.

4.9 In case, the farmer farmer/lessee farmer/lessee farming organization /operator is not satisfied with the observations of the Agriculture/Horticulture Officer in Columns 15 and 16 in Form – 4(B), he may, if he so desires, prefer an appeal to the higher authorities in the concerned department of the State Govt.

4.10 Each concerned Agri./Hort. Officer will examine the certificate of residue analysis and suggest control/corrective measures in terms of do's and don'ts.

4.11 At the end of the pomegranates season, the Agriculture/Horticulture Officers will have the responsibility to examine the consolidated test reports submitted by the NRL to the State Government and also the copies e-mailed by the laboratories to them. They will also suggest corrective action to the farmers (please see para 9.7).

		4.12	Agriculture/Horticulture Officers will organize meetings with farmers/lessee farmer/lessee farming organisation and exporters regularly to provide guidance on the quality production of pomegranates.
		4.13	Agriculture/Horticulture Officers shall carry out awareness programmes to allow use of only registered/recommended agrochemicals for pomegranates as given in Annexure-5.
		4.14	The Agri./Hort. Officers will also provide information on active ingredients of pesticide products available in the market based on products tested by the concerned Govt. institutions.
		4.15	Agriculture/Horticulture Officers will ensure that only those agrochemicals/pesticides are allowed to be used by the farmers/ lessee farmer/lessee farming organisation, which have complete information on labels including generic name, spray schedule, dosage, name of pest/disease affected, waiting period and also have an accompanying leaflet in local language and in English.
		4.16	In the event of alert information issued by the NRL in respect of a pomegranates farm (please see para 9.11), the State Government will suspend export of pomegranates from that farm/plot until the alert notice is revoked by the NRL based on re-testing (please see para 9.12).
5.	Procedure for sampling from pomegranates farms/plots	5.1	Farmers/lessee farmers/lessee farming organisations/ exporters should provide a schedule to laboratories well in advance for drawl of samples to enable them to plan their sampling arrangements.
		5.2	Samples of pomegranates intended for exports shall be drawn for testing by the nominated laboratories and NRL listed in Annexure-6 of this procedure.
		5.3	Samples for laboratory testing shall be drawn by a representative of lab authorized for this purpose. At the time of sampling, he will obtain from the farmer, copy of the Annexure 2. In case, if there is a time gap of more than 30 days between sampling and final harvesting, then new sample shall be drawn from the plot.
		5.4	After drawl of samples, the lab official shall record the quantity of sample drawn and place his signature and date at the back of Registration Certificate.
		5.5	Laboratories are advised to enter details of samples drawn from the registered plots in the web-based software placed on the Apeda website immediately after drawl of samples. The Plot Registration Number/Numbers will remain the key numbers for all purposes.
		5.6	The samples of pomegranates, soil and water shall be drawn in the presence of farmer/lessee farmer/representative of lessee farming organisation and exporter as per the procedure given in Annexure–7.

5.7 Each sample will be drawn from the concerned plot/ plots/ farm, packed separately in two corrugated cartons [one will be the lab. sample and the other will be the counter sample (to be retained for test by NRL in case of dispute)]. The cartons will be sealed and signed, and handed over/ transferred to the residue – testing laboratory within 24 hours along with the sample slip as per format given in Annexure-8. The sample slip shall be signed by the farmer/lessee farmer/representative of lessee farming organisation, exporter and representative of the laboratory who has drawn the sample. At the time of signing the certificate in this annexure, he will also ensure that a copy of the Annexure-2 has been obtained. The lab will not be able to proceed forward with entries in the software unless these documents have been duly obtained, verified and entered into the software.

5.8 It is recommended that the laboratories check both the forms under Annexure-4, i.e., Annexure- 4(A) and 4(B) for a better understanding of the use of chemicals on the plot. The lab representative will also verify the signature of the farmer or his authorised representative on the sample slip with those given on Annexure- 4(A) and 4(B).

5.9 Laboratories shall not draw samples plot(s) under the following conditions :

a) If the plot(s) is/are not actually registered with the Distt. Agri./Hort. Office or the plot(s) is/are under suspension/have been cancelled for export to the E U.

b) If the Registration No. of the plot(s) is not in the coding format given in para 3.7 above,

c) If the farmer/exporter fails to provide a copy of the Registration Certificate along with sketch/ drawing of the lay- out of the adjoining plots (please see paras 3.5 and 3.6 above),

d) In case of any inconsistency (10% variation allowed) in the calculation of area of plot/plots/farm and likely production stated in the Annexures-2 (for area purposes) and 4(B),

e) Unless there is a clear recommendation by the Agri/Hort Officers in Column 15 of the Annexure-4(B) to this effect,

f) If the signature of the farmer or his authorised representative on the sample slip does not match with those given on the Annexure-3 and 4(A) & 4(B),

g) Until the laboratory representative has verified that the sample is being drawn from the plot/plots/farm whose registration is valid,

h) Until relevant calculations have been carried out as per procedure given in Annexure-7, and

i) Unless the sample slip, duly filled, accompanies the sample. The samples and the sample slip will go together to the laboratory.

5.10 For movement of the samples from farm to the laboratories, a record showing the chain of custody in the following format will be maintained, and this record will also move with the samples and documents :

Person	Name	Date	Sign
Sampler			
Courier-1			
Courier-2			
Laboratory			

5.11 The representative of the laboratory drawing the sample shall mark the sample with the sample slip number so as to co-relate the sample slip with the sample drawn.

5.12 On arrival in the laboratory, each sample shall be numbered by indicating the relevant code number as per procedure followed by the laboratory.

5.13 All the exporters/recognised pack houses and laboratories shall retain representative samples (see para 8.4) of the exported pomegranate in their cold storages for a period of 90 days from the date of test report as per directions given in Annexure – 7.

5.14 The representative of the laboratory drawing the sample shall mark the sample with the sample slip number so as to co-relate the sample slip with the sample drawn.

6. Accreditation/ recog nitions requirements and responsibility of authorized laboratories

6.1 The authorized laboratories shall be accredited to the National Accreditation Board for Testing and Calibration Laboratories (NABL) to ISO/IEC-17025.

6.2 The authorized laboratories shall have a valid APEDA recognition under its scheme for laboratory recognition.

6.3 The authorized laboratories shall carry out sampling as per method of sampling given in Annexure-7 and shall, at the request of the concerned farmer/exporter, test the pomegranates for residue levels of the chemicals, a recommended list of which is given in Annexure-9. Annexure-9 also consist names of chemicals banned in India or EU.

6.4 Based on the findings of the previous season for the banned chemicals, soil and water testing shall be done from farm areas identified by the NRL. The test report should reflect the detection limits and results. Copy of report shall be provided by the laboratories to the State Government/Farmer/exporter/ Registration Authority and NRL.

6.5 After ensuring that para 5.7 to 5.9 are complied with, the recognised laboratory shall nominate its representative for drawing samples of pomegranates, soil and water as per the procedure given in Annexure-7.

6.6 The certificate of residue analysis shall be issued as per the format given in Annexure-10 only through the AnarNet with digital signature of the representative of Laboratory. The residue analysis certificate must be issued within six days of the drawl of samples. This includes the day of sampling also. The authorized laboratory would issue residue analysis certificate of detected agro chemicals, however, upload in AnarNet all the agrochemicals under monitoring.

6.7 The laboratory shall calculate the area of the farm/plot(s) [Sr. No. 5] and total likely production of the farm/plot(s) [Sr. No. 6] declared in Annexure-10 on the basis of Annexure-2 (for area purposes) and Annexure-4(B) only. Inconsistency beyond the 10% variation, if any, observed in the two documents shall be immediately reported by them to the NRL/State Government [please see paras 5.9 (d) and 8.13].

6.8 In case, the test results exceed the MRLs with respect to the destination of the consignment declared by the exporter/ farmer/ lessee farmer/lessee farming organisation, the nominated laboratory will immediately (within 24 hours) bring the matter to the notice of NRL, Horticulture/Agriculture officer (whose address is given in Annexure – 2) and APEDA along with a copy of the test report giving details of the plots and the agrochemicals exceeding the levels. The laboratories shall, in case of failed samples, also send the chromatograms, etc. to the NRL by speed- post/courier.

6.9 In case a Pomegranate sample fails and an internal alert has been issued by the NRL, the farmer/lessee farmer/lessee farming organisation /exporter may choose to have re-sampling done at a later date (see Annexure-8). In such cases, if the second sample passes the test, the laboratory shall, without delay, intimate the NRL to enable them to revoke the Internal Alert Information, which shall be effective from that date (see para 9.12).

6.10 The inspection of the authorized laboratories may be carried out by NRL/PSC issuing Authority without prior notice to the laboratory (see para 9.4).

6.11 The authorized laboratories shall participate in the training/inter-laboratory proficiency testing organized by the NRL. The laboratories shall inform in writing to APEDA and NRL before commencement of the sampling and analysis of pomegranates regarding their competence and readiness.

6.12 The authorized laboratories shall not add any additional statement/ disclaimer with regard to sampling, analysis and grading of pomegranates meant for exports.

6.13 In case the authorized laboratories observe or arrive at an impression during the course of the analysis that there is likelihood of the presence of a chemical not listed in Annexure-9, the same shall be immediately (within 24 hours) brought to the notice of NRL/Sate Government/Inspecting Authority and Horticulture/ Agriculture Officer alongwith, where possible, a copy of test report.

7.	Procedure for issue of Certificate of AGMARK Grading (CAG) and Phyto Sanitary Certificate (PSC)	7.1	The complete procedure for grant of Certificate of Authorization and the Certificate of AGMARK Grading (CAG) is set out in Annexure-11. The CAG shall be issued only after receipt of the inspection report from the laboratories through AnarNet. The exporters shall ensure availability of the chemicals, apparatus, etc. as given in Appendix-(i) to the Annexure-11.
		7.2	The farmer/exporter shall request one of the Government of India notified under Quarantine Regulation (IPPC 1951) to State PSC Authorities to issue the Phyto Sanitary Certificate along with application prescribed annexure with other necessary documents as prescribed by Plant Protection Advisor, Government of India or their nominated laboratories to issue the Phyto Sanitary Certificate alongwith the Exporter's/Shipper's declaration given in Annexure-12 electronically with scanned copy of fumigation certificate, if any and the container loading sheet/packing list/ proforma invoice/copy of LC. The laboratory shall verify the following documents at the time of inspection : a) Packhouse Recognition Certificate issued by APEDA, and b) Certificate of Authorization issued by DMI. c) Fumigation certificate for wooden packing material issued by the Government of India accredited MBR fumigator as per NSPM-12. d) AGMARK Grading certificate issued by DMI. e) Copy of Contract/LC for additional declaration regarding quality, quarantine issues and pest and diseases to be given in the Phytosanitary certificate to fulfil the quarantine regulations of importing country.
		7.3	The onus of proving veracity of the contents of the Annexure-12 lies with the exporter/shipper.
		7.4	The PSC Authority shall issue the PSC only after fulfilments of procedure laid down by Plant Protection Advisor to Govt. of India as per International Plant Protection Convention (IPPC) and following Guidelines and after satisfying themselves about following conditions: (I) signature of farmer/lessee farmer/lessee farming organisation on the sample slip matches with those given on Annexures 3, 4(A) and 4(B).
		7.5	The PSC issuing authorities shall, on random basis, at the packhouse, draw a representative sample from the consignment, which shall be sealed/marked properly and handed over to the exporter/packhouse for storage (please see para 5.12).
		7.6	The PSC official shall mark the representative sample drawn with the Laboratory Test Report Number(s).
		7.7	The container loading sheet (packing list) should contain details of farmer name, farmer code, quantity per packet, total quantity, etc.
		7.8	The exporters/recognised packhouses and laboratories shall retain representative samples (see para 7.5) of the exported pomegranates

			in their cold storages temperature range of 5-10°C and Relative Humidity of 90-95% for a period of 60 days from the date of test report as per directions given in Annexure-7.
		7.9	Agmark and Phyto Sanitary Certificate issuing authority shall educate the farmers, exporters and other stakeholders regarding the requirements of grading and quarantine related issues of the importing countries.
8.	Responsibilities of the National Referral Laboratory (NRL) at National Research Centre for Grapes (NRCG)	8.1	The NRL shall prepare the recommended list of chemicals to be used for cultivation of pomegranates given in Annexure-5 before commencement of the cultivation season every year.
		8.2	The NRL shall submit to APEDA and State Governments proposal of the updated list of chemicals recommended for the control of various diseases and insect pests and a dynamic list of chemicals and any other contaminants to be analyzed for the pomegranates with their MRLs by June end which shall be finalized in consultation with the exporters and farmers.
		8.3	The NRL shall specify and verify method of sampling and analysis to the APEDA recognized laboratories authorized for sampling and analysis. The NRL shall make recommendations to APEDA for authorization of the laboratories for sampling, analysis of pomegranates for exports.
		8.4	The NRL shall prescribe the list of chemicals and their MRLs for the purpose of testing before commencement of pomegranate export season based on prescription for cultivation of pomegranates as well as EU list of MRLs of chemicals including the list of banned chemicals for exports of pomegranates. This list of chemicals to be tested for their MRLs shall be revised regularly.
		8.5	The NRL shall monitor the work of authorized laboratories by conducting surveillance audit to ascertain that they are following the criteria.
		8.6	The NRL shall compile residue analysis data of the authorized laboratories for each year. On the basis of the data, the NRL shall also prepare a plan of action for the following year.
		8.7	The NRL shall draw 5% of the samples directly from the recognised packhouses pertaining to the batches tested by the designated laboratories as a measure of conformity. The NRL shall analyze these samples and integrate the reports in the consolidated report.
		8.8	The NRL shall also evaluate 5% test data of the samples analyzed by the authorized laboratories pertaining to the batches tested by the authorized laboratories as a measure of conformity.
		8.9	Where residue levels are found to be higher than permitted levels, depending upon the destination of the consignment declared by the exporter/ farmer, the NRL shall advise the exporters/farmers about the control measures to be taken.

8.10 The NRL shall evaluate the soil and water test reports to be obtained by the NRL from the authorized laboratories of the samples analyzed from the farms where banned chemical was detected in the previous season as well as on the basis of the information contained in the test reports of the current season expeditiously and suitably advise the concerned farmer/exporter and the Registration Authority (Agriculture/Horticulture Officer).

8.11 The NRL shall bring immediately to the notice of farmers, growers, exporters, APEDA, State Governments and other stake holders regarding exceeding levels of chemicals that has been recommended by the NRL for plant protection and cultivation of pomegranates.

8.12 The NRL shall organize training on testing of each residue or groups of residues for the authorized laboratories.

8.13 NRL shall organize inter-laboratory/proficiency testing 2 times during the pomegranate season and guide the laboratories.

8.14 The NRL shall organize interactive meetings of stakeholders on a regular basis. Participants in these meetings shall include farmers, exporters, CIB&RC, Agriculture/Horticulture Officers, PSC issuing authorities, Agmark officials, ICAR, any other service providers such as agro chemical producers and suppliers and APEDA.

8.15 The NRL shall update itself, APEDA, State Governments, farmers and producers, exporters and the authorized laboratories with regard to the list of chemicals with their MRLs and method of sampling and analysis.

8.16 Upon receipt of an alert notice from a authorized laboratory about a failed sample, the NRL shall, without delay, and in any case within 24 hours unless the NRL considers it necessary to carry out any investigation, issue an Internal Alert Information to the State Government, exporters, PSC issuing authorities, Laboratories and APEDA under intimation to the farmer in respect of the farms in case of detection of higher residues or major elements than the limits prescribed in this document as amended from time to time. A format of Internal Alert Information is given in Annexure - 13.

8.17 In case of Pomegranate plot/plots/farm having same package of practices and where the farmer/lessee farmer/lessee farming organisation gives an undertaking about same package of practices being followed, only one random sample shall be drawn from the consignment meant for export, which will be sealed/marked properly and handed over to the exporter/pack house for storage. In case a Pomegranate plot/plots/farm, on re- testing of a sample, passes the MRL test (see para 6.9), the NRL shall, without delay, revoke the Internal Alert Information, which shall take effect on that date. In this regard, the NRL shall intimate all concerned about new status of the plot.

		8.18	In the event of any report of inconsistency received from a laboratory in terms of para 6.7 above, the NRL shall immediately take corrective action through interaction with the State Government (Agriculture/ Horticulture offices).
9.	Powers of National Referral Laboratory	9.1	The NRL shall have the right to draw samples from registered pomegranate farms, packhouses and laboratories.
		9.2	The NRL shall have the right to verify analysis data corresponding to the samples drawn and/or tested by the designated laboratories.
		9.3	The NRL shall have the authority to recommend to APEDA and/ or NABL, derecognition of authorized laboratories in the event of non-compliance with the method of sampling and analysis for pomegranates.
		9.4	The NRL shall have the authority to inspect the authorized laboratories without prior notice (see para 6.10).
10.	Responsibilities of farmers, growers and exporters	10.1	The farmers/growers/exporters and any other stake holders of pomegranates shall comply with the MRLs of chemicals based on importing country's Regulations. Compliance with regulations on pomegranates such as grade, quality, safety and wholesomeness of pomegranates, pre and post harvest practices, plant and its quarantine, packing, fumigation, certification for wooden pallets, sanitary and Phytosanitary measures, any other requirements shall be the responsibility of the farmer, growers, exporters and other stake holders for exports of pomegranates.
		10.2	The farmers/growers/exporters shall be under obligation to apply only those chemicals as are recommended by NRC Pomegranates (Annexure-5). Use of spurious and contaminated chemicals and any other agri inputs for production of pomegranates shall be at the sole risk and cost of the farmers/growers/ exporters.
		10.3	The farmers/growers/exporters shall clearly inform to the PSC issuing Authority/Agriculture/Horticulture Officers regarding compliance with Good Agriculture Practices (GAP) quality/food safety management systems and/or any other compliance certification and inspection systems implemented and carried out for production, processing and exports of pomegranates as per intended use of these systems including certification and inspection carried out by the concerned agencies.
		10.4	The farmers/growers/exporters and other stake holders of pomegranates shall provide to NRC Pomegranates, the list of chemicals to be tested for exports of pomegranates. On the basis of feedback provided by the farmers, growers, exporters and other stake holders the NRL shall recommend for testing of other chemicals in consultation with APEDA.
		10.5	Samples of soil and water from the registered farms containing banned/restricted chemicals in the previous season shall be drawn for testing by the authorized laboratories and the NRL.

		10.6	While a recommended list of chemicals to be tested is given in Annexure-9, farmer/exporter shall have responsibility to check if any additional chemical is required to be tested by the authorized laboratories with respect to the farm from where the exporter is sourcing the produce based on the spray records maintained by the farmer in Annexure-2 and signed by the Inspection Authority (Agriculture/Horticulture Officer).
		10.7	The farmers/growers/exporters and any other stake holders shall have the responsibility to update and inform the list of chemicals and any other contaminants to be tested and monitored for export of pomegranates through their trade intelligence information.
		10.8	In the event of any non-compliance of the importing country's Regulations by the farmers, growers, exporters and other stake holders, the liability of losses shall remain with the farmers, growers, exporters and other stake holders.
		10.9	The farmers/growers/exporters and other stake holders shall ensure correct measure of pomegranates in their respective packing on arrival in the market until the pomegranates are lifted by the consumers from the retail market shelves.
11.	Overall monitoring and responsibilities of APEDA	11.1	APEDA shall be facilitating export promotion of pomegranates as envisaged in the APEDA Act.
		11.2	APEDA may inform the Governments of importing countries and the European Commission the names, addresses of recognized packhouses as well as the designated laboratories/PSC issuing authorities notified by the Government of India and will also display the list on its website quarantine related issues of the importing countries as obtained from the Ministry of Agriculture.
		11.3	APEDA will, through the NRL, regularly monitor the functioning of each authorized laboratory to ensure implementation of the procedures laid down in these guidelines based on its testing capacity for chemicals residue analysis.
		11.4	APEDA will evaluate the weekly test results submitted by the authorized laboratories and will require that the control measures suggested by the NRL be implemented by the State Government or laboratory, as applicable.
		11.5	Where necessary, APEDA will nominate a Committee consisting of the representatives of exporters/association, designated laboratories and APEDA under the leadership of National Referral Laboratory. As a test, the committee will evaluate the procedure followed by the exporters. A sample size of 5% covering both small and large packhouses will be taken for the purpose. Complete records of 5% quantity of pomegranates, taken on a random basis, exported by the exporter will be checked at the end of the season.
		11.6	APEDA will assess the work carried out by the NRL with respect to the responsibilities laid down in this document as amended. APEDA shall provide RMP document to the respective State Governments before start of the season for effective monitoring of traceability for exports of pomegranates.

12.	Penal Provisions	12.1	In the event of breach of these procedures for controlling agro chemical residues in pomegranates, APEDA may initiate action as per the provisions of section 19(3), Chapter-V of APEDA Act, 1985, in addition to the following: a) Cancellation of the Registration-cum-Membership Certificate of exporters. b) Derecognition of packhouses. c) Notifying to DGFT for cancellation of Import-Export Code No. allocated to such exporters. d) Any other action as deemed fit.

Annexure-1 : Physical Document

Application for Registration/Renewal of Pomegranate Plot(s) for export
(To be submitted by the farmer lessee farmer/lessee farming organisation)

To

Registration Authority & District Superintending Agriculture Officer

Taluka—————— District —————— State ——————

Sub: Registration/Renewal of Pomegranate Farm for Export

Dear Sir,

You are requested to kindly register / renew my pomegranate farm for export. Other necessary details are as follows:

1. Full name of the Pomegranates growers
   Father's /Husband's name
   Name of partners

a. Correspondence address
   At Post
   Taluka
   District
   State
   Telephone No with STD code no.
   Mobile No.
   E-mail address

b. Farm/Plot location address (Survey No/ Plot No.) along with map/layout of the plot with indication of all sides of crop grown.
   (please attach copy of 7/12)

2. Pomegranates Farm registration No.
   (In case of renewal of garden)

3. Total Farm area (in Ha)

4. Pomegranates farm is certified with
Global GAP if yes give details
(attach copy)
Certificate No.
Date of issue
Date of validity
Name of certification agency

5. Number of plots in the farm with area of each plot

Plot no.	Area (inHa)	Survey/ plot No	Variety	Date of plantation	Date of pruning	Likely production (MT)
Plot no.01						
Plot no.02						
Plot no. 03						

6. Probable date of harvesting

7. Pack –house registration number, if any and its validity

8. Application fee of Rs. 50/-per plot / year
Challan no
Name of treasury
Date of payment deposited in treasury

9. Details about last year export
Quantity in MT
Name of Exporter
Name of Pack house
Name of Laboratory where sample was analyzed

10. Whether Internal Alert Notice issued by NRL, Pune in previous season (In case of renewal of pomegranates farm give details)

It is certified that the information mentioned above is correct and the plot mentioned above is not under suspension / has been cancelled for export of pomegranates.

Date : Signature of the Farmer

Place : Name of the Farmer

Annexure-2A Physical Document

Plot registration and field chemicals application record (to be maintained by the farmer/exporter)

(Copy to be given to authorized representative of the laboratory at the time of sampling)

1) Plot Registration Number :
2) Date of Registration/Renewal of plot :
3) Name of the farmer/operator with address :
4) Location of plot (lay out/benchmark) * :
5) Total area of the registered farm/plot (Ha) :
6) Name of pomegranates variety :
7) Date of planting :
8) Date of pruning :
9) Likely production :
10) Chemicals application machinery used :
11) Pack-house registration no, if any, and its validity :
12) Technical authorization for use :

Sr. No.	Date	Day	Trade name of	Active ingredient	Batch	Target	Preventive	Agri chemical	**Agro chemical	** Water	Pre-harvest	Days Estimate	Time Actual	Name & signature
1	2	3	4	5	6	7	8	9	10	11	12	13	14	15

* The drawing/sketch of the layout of the farm / plot(s) also showing the adjoining properties must be enclosed.

** Plot size constitutes a maximum of 1.2 Ha (see para 3.9).

1. It is certified that registration of the above mentioned plot has been done as per procedure given in para 3.2 to 3.11 of the "Procedures for Export of Pomegranates" and details of the above plot have been entered on the AnarNet maintained on the APEDA web-site.
2. It is also certified that a copy of the annexure along with a copy of the Registration Certificate, drawing/map layout of the plot and this annexure, duly signed, have been handed over to the farmer/exporter as the case may be.

The above information is correct. The total likely production of this plot during previous season is estimated to be … MT. Checked and certified by Agriculture/Horticulture Officer Name and Signature:

Official address of Agriculture/Horticulture Officer (mandatory)

Date: Signature of the Farmer/Authorized Operator

Place : Address (mandatory)

Annexure-2B Physical Document

Government of_____
Department of_____
Certificate of Registration of Pomegranate Farm for Export

This is to certify that__________is here by registered as Pomegranates Grower/Pomegranates Exporter with the office of the District Superintending Agriculture/ Horticulture officer,__________in accordance with the Procedure for export of pomegranates.

The detail of the registered Pomegranates Grower is as follows:

Name of the Pomegranates Grower

Full Address Village

Taluk/Mandal District

Sr. No.	Survey/GAT No.	Plot No.	Variety	Area of Plot (Ha.)	Farm Reg. No.

1) Map Layout enclosed.
2) This certificate is valid up to
3) Have verified the Survey/GAT No. with respect to the registration and to the best of my knowledge the above information is correct.

Place:
Date:

Registration Authority and District Superintending Agriculture/ Horticulture Officer

Terms & Conditions of Registration of the Pomegranates farm for Export)

a) To follow only the recommended package of practices)
b) The Farmer shall not use agrochemicals other than those allowed for use on pomegranates and listed in Annexure-5

c) Misbranded, Non-recommended or banned agrochemicals or any harmful chemical shall not be used

d) After drawl of sample for residue testing, spraying/application of any pesticide shall not be carried out

e) The registered farmer shall maintain record of package of practices followed by them in a prescribed register to be prescribed by this office

f) The registered farmer/exporter at the time of harvest shall give a declaration in annexure-5 stating that no pesticide, insecticide, weedicides etc. have been sprayed/applied after drawl of the sample for laboratory analysis

g) No amendments will be made by them on the registration records or the registration certificate).

Annexure-3 - Physical Document

Declaration

(To be given by the farmer to the exporter)

I,__________, resident of__________and having pomegranates farm with Plot(s) Registration Nos.__________, renewed on__________(if applicable), hereby, certify that :

1) All plots under my/our farm (including those that are not intended to be utilized for export purposes) have been registered with the District Agriculture/Horticulture Office by following the procedure laid down in this document and that none of the plots mentioned above are under suspension/have been cancelled for export to the E U.

2) On.__________, I have allowed drawl of the pomegranates samples by the authorized representative of.__________(laboratory) for testing. After drawl of samples as per procedure prescribed in Annexure-8 of the Procedures for Export of pomegranates, I have not sprayed any kind of chemicals or contaminants, insecticides, fungicides, weedicides, including herbal products (including growth regulators) on the pomegranates plot other than those recommended by NRC Pomegranates.

3) Harvesting of__________MTs/Kgs. of pomegranates from my plot(s) has been carried out under my supervision on__________and the pomegranates have been stored in__________numbers of crates/ boxes/ etc.

4) I propose to effect export of the harvested pomegranates myself/through M/s (exporter). The address of cold storage/packhouse shall be as follows :

   __________(APEDA Registration No. of pack house and its validity)

5) The balance quantity of approximately__________MTs/Kgs. of pomegranates remaining in the plot(s) referred to above shall be informed to the District Agriculture/Horticulture Office.

Date :

Place :

Signature of Farmer/
Authorized person by the farmer with
Plot Registration Numbers

Annexure-4 (A) Electronic Document

Inspection report of pomegranates farm/plot to be maintained by the Inspecting Authority (Agriculture/Horticulture Officer) and farmer first inspection (At the time of new registration/ renewal of pomegranates garden for export)

1. Name and address of the Farmer / grower
   At. Post
   Taluka
   District
   State
   Phone no. with STD code
   Mobile no.
   E-mail address
2. Plot Registration No and date of renewal
3. Address of the Plot
   Survey No. / plot no.
   At. Post
   Taluka
   District
   State
4. Total area of the Plot Map of the Plot (Please indicate all sides of farm crop grown and attach Annexure -2 and Map layout)
5. If late registration indicate areas demarcated
6. Whether Plot is certified for Good Agriculture Practices (GAP) if so attach a copy of valid certificate
   Certificate No.
   Date of issue and validity
7. Last year's export details
   Quantity (MT)
   Name of the Exporter
   Name of Pack house
   Name of Laboratory where sample was analyzed

8 Whether Internal Alert Notice issued by NRL, Pune during the previous season
  Yes/No (if yes give details of the alert notice)

9. Date of flowering

10. Condition of crop relating to pests & diseases & stage of crop

11 Any advice given to the farmer

12 Recommendations of Inspecting authority (Whether plot is fit for registration / renewal of registration)

Yes/ No (If no give specific reason)

13. Date of Inspection

It is certified that the registration details of the above plot has been entered on the web based database maintained on the APEDA web-site

Signature of farmer/Grower	Signature of Inspecting Authority
Name of Grower	Full Name of Inspecting Authority
	& full address with office seal

Annexure-4 (B) Electronic Document

Inspection report of exportable registered pomegranates farm/ plot to be maintained by the Inspecting Authority (Agriculture/Horticulture Officer) and farmer second and final inspection report for export (copy must be given to representative of laboratory at the time of sampling)

1. Farm/Plot Registration No.

2. Name and Address of the Pomegranates Grower
   Full Name
   At. /Post
   Taluka
   District
   State
   Phone No/Mobile No.

3. Address of the Farm/Plot
   Survey No/ Gat No
   At. Post
   Taluka
   District
   State

4. GAP Certificate No. if any

5. Total area of the plot (Ha)

6. First Inspection Report No. and Date

7. Condition of the crop relating to pest and diseases and quality of the crop

a)	Bacterial blight	Yes/No
b)	Wilt	Yes /No
c)	Leaf Fruit Sports	Yes/No
d)	Fungal Blight	Yes/ No
e)	Fruit Borer	Yes/No
f)	Stem Borer	Yes/No
g)	Mealybug	Yes/No
h)	Thrips/Aphids/Jassids/White flies	Yes/No
i)	Mites	Yes/No
j)	Nematodes	Yes/No
k)	Weedicides	Yes/No

	i) Any other observation concerning quality (browning etc)	Yes/No
8.	Verification of spray records with respect to list of agrochemicals and other agri inputs as recommended by NRC Pomegranates for control of various diseases and insect pests	Yes/ No.(if no please substantiate)
9.	Likely total harvest of the plot (MT) (No of plants in Ha and average no of plant/average weight of fruit)	
10.	Tentative date of harvesting	
11.	Name of residue laboratory where samples being analyzed	
12.	Whether agro chemicals, plant growth regulators and other chemicals spraying schedule has been as per the NRC Pomegranates recommendation	Yes/No(if no please substantiate)
13.	Whether farmer has followed other advice/recommendation of Agriculture/Horticulture Officer during the year	Yes/No(if no please substantiate)
14.	Whether sampling should be done by the Residue Laboratory (please see para 4.7 of procedure)	Recommended / Not recommended
15.	Reason for not recommending drawl of samples (please see para 4.7 of procedures)	
16.	Advice/Recommendation given to the farmer concerning incidence of pests and diseases and quality of pomegranates at this stage	
17.	Date of Inspection	

Signature of Pomegranates Grower
Name of Grower

Signature of Inspecting Officer
Name and Full address of Inspecting Authority with seal

CC: 1. Registration Authority (Pomegranates Farm/Plot)

2. Representative of Residue Laboratory

Endorsement by the sample drawing laboratory

This is to certify that I have personally drawn the samples of pomegranates from this plot for the purposes of laboratory analysis and by adopting the procedure given in Annexure-7. I have obtained a copy of currently valid Registration Certificate plot drawn/map layout and that the location of this plot is as per the map. I have verified that the registration of the plots is/are valid. I have also obtained a copy of Annexure-2.

Date : Signature

Place: Name of authorized representative of Nominated Laboratory with office address

Annexure-5

Agrochemicals recommended for control of various diseases and insect pests for export of Pomegranate

Sr. No.	Pesticide recommended for major disease	Nature of Pesticide	Dose on formulation basis	EU MRL (mg/kg)	Pre-harvest Interval (PHI in days)
DISEASES					
A.	**Bacterial blight** *(Xanthomonas axonopodis* pv. *punicae)*				
1.	Streptomycin Sulpihate 90% + Tetracycline hydrochloride 10%	S	0.5g/l	0.01*	55
2.	Copper compounds (including Coppe roxychloride 50% WP, Copper hydroxide 77% WP *etc.*)	NS	2-2.5g/l	20	60
B.	**Wilt** (Fungal complex *Ceratocystis fimbriatam Fusarium oxysporum)*				
3.	Propiconazole 25% EC	S	1.50 ml/l (drenching)	0.05	20
4.	Carbendazim 50% WP	S	2.00 g/l	0.1	100
5.	Tridemorph 80% EC	S	1.0 ml/l 2.0 (drenching)	0.01	40
C.	**Leaf Fruit Spots** *(Alternaria alternata, Cercospora punicae, Colletotrichum* sp., *Drechslera* sp., *Sphaceloma* sp.*)*				
6.	Copper compounds (including Copper oxychloride 50% WP, Copper hydroxide 77% WP *etc.*)	NS	2-2.5 g/l	20	60
7.	Mancozeb 75% WP	NS	2.0 g/l	0.05	90
8.	*Propineb 70% WP	NS	3.0g/l	0.05	90
9.	Copper hydroxide 77% WP	NS	2.0 g/l	20.0	60
10.	Ziram 80% W	NS	2.0g/l	0.05	90
11.	Captan 50% WP	NS	2.5g/l	0.02	35

12.	Chlorothalonil 75% WP	NS	2.0 g/l	0.01	90
13.	Difenoconazole 25% EC	S	1.0 g/l	0.1	90
14.	Triadimefon 25% WP	S	0.5-1.0 g/l	0.1	40
15.	Sulphur 80% WP	NS	2.5 g/l	50	15
16.	Carbendazim 50% WP	S	1.0 g/l	0.1	90
17.	Thiophanate Methyl 70% WP	S	1.0 g/l	0.1	50
D.	**Fungal Blight** *(Phytophthora* sp*)*				
18.	Mancozeb 75% WP	NS	2.0g/l	0.05	90
19.	Copper Compounds (including Copper oxychloride 50% WP, Copper hydroxide 77% WP *etc*)	NS	2-2.5 g/l	20	60
20.	Metalaxyl 8% + Mancozeb 64% (Metalaxyl MZ 72% WP)	S	2.5 g/l	0.05+ 0.05	90
21.	Cymoxanil 8%+Mancozeb 64% (Curzate M8)	S	2.0 g/l	0.05+ 0.05	90
22.	Fosetyl- AI 80%WP	S	2.0 g/l	75	30
23.	Dimethomorph 50% WP	S	1.0 g/l	0.05	66
24.	Azoxystrobin 23 SC	S	0.5-1.0 ml/l	0.05	45
25.	Pyraclostrobin 20%	S	1.5 kg/ha	0.02	60
INSECT AND OTHER PESTS					
E.	**Fruit Borer** (*Deudorix isocrates*)				
26.	Indoxacarb 14.5% SC	NS	0.5 ml/l	0.02	30
27.	Spinosad 45% SC	NS	0.5 ml/l	0.02	28
28.	Cypermethrin 25% EC	NS	1.0 ml/l	0.05	40
F.	**Stem Borer** (*Celosterna spinator*), **shot hole borer** (*Xyleborus fernicatus*)				
29.	Chlorpyriphos 20% EC	NS	2.0 ml/l	0.05	40
30.	Indoxacarb 14.5% SC	NS	0.5 ml/l	0.02	30
31.	Spinosad 45% SC	NS	0.5 ml/l	0.02	28
32.	Cypermerthrin 25%EC	NS	1.0 ml/l	0.05	40
G.	**Mealy bug** (*Ferrisia virgata*)				
33.	Chlorpyriphos 20% EC	NS	2.0 ml/l	0.05	40
34.	Dimethoate 30% EC	S	1.0 ml/l	0.02	100
35.	Imidacloprid 17.8% SL	S	0.3 ml/l	0.05	90

36.	Thiamethoxam 25%WG	S	0.25 g/l	0.05	40
37.	Methomy1 40 SP	S	1.0g/l	0.02	90
H.	**Thrips/Aphids/Jassids/White flies**				
38.	Dimethoate 30% EC	S	1.0 mL/l	0.02	100
39.	Imidacloprid 17.8% SL	S	0.3 mL/l	0.05	90
40.	Acetamiprid 20 SP	S	0.3 mL/l	0.01	90
41.	Thiamethoxam 25%WG	S	0.25 g/l	0.05	40
42.	Lambda-Cyhalothrin 05 CS	NS	0.20-0.5 ml/l	0.02	80
43.	*Cyantraniliprole 10.26% OD	S	0.7-0.9 ml/l	0.02	90
I.	**Mites**				
44.	Propargite 57% EC	NS	1.0 ml/l	0.01	15
45.	Abamectin 1.9% EC	S	0.5 ml/l	0.01	30
46.	Azadirachtin 1%	NS	2.0 ml/l	0.01	3
J.	**Nematodes**				
47.	Azadirachtin 1%	NS	2.0 ml/l	0.01	3
48.	Phorate 10 G	S	25g/plant	0.01	Data not available
49.	Carbofuran 3G	S	40g/plant	0.01	Data not available
K.	**WEEDS**				
50.					

NS= Non systematic, S= Systemic

Source: ICAR-NRC on Pomegranate, India

Note

- Agrochemicals with asterisk (*) with label claim rest without label claim, hence recommendation is Adhoc.
- As the data based on scientific field trials on PHI for pomegranate are not available for most chemical hence, PHI given are only indicative and adhoc in nature as are based on PHI for other fruit crops grown in similar climatic conditions and residue analysis of limited samples of harvested produce in past years, hence, may change at later stage on availability of scientific data and are of advisory nature and therefore, not covered under any legal scrutiny.
- Recommended agrochemicals for the management of various insect pests and diseases along with their dose, PHI and MRL values are recommendations by SAUs, ICAR Institutes & research literature and of

advisory nature for the Good Agriculture Practices and therefore, not covered under any legal scrutiny.

- All the doses mentioned above are for high volume sprayers, where normal spray volume is 800-1000 L/ha. Spray volume can however be changed as per the efficiency of sprayers used. However, the amount of each pesticide (active ingredient) recommended for 1 ha on the basis of 1000 L spray solution should be strictly maintained to minimize pesticide residues.

Annexure–6

List of authorized laboratories for Anar Net

No.	Name and contact details of the laboratory	Scope
	National Research Centre on Grapes (Indian Council of Agricultural Research) P.B. No. 3, Manjri Farm Post, Solapur Road, Pune 412 307 Tel.: +91-20-26956002 EPABX: +91-20-26956000 Fax: +91-20-26956099 nrcgrapes@gmail.com; apedanrl@gmail.com;	NRL for products of plant origin & NABL
1.	Centre for Food Testing, Bharati Vidyapeeth Deemed University Centre for Advanced Research in Pharmaceutical Sciences Building 5th Floor Erandwane Paud Road Erandwane Pune 411 038 Tel: 020-65737381,82,83 cft.bvdu@gmail.com;	Recognized by APEDA & NABL accredited
2.	Envirocare Labs Pvt. Ltd. A-7 MIDC Wagle Industrial Estate Main Road Thane 400 604 Tel: 022-25838286-88 Fax: 25838289 info@envirocare.co.in;	-do-
3.	First Source Laboratory Solutions LLP (Analytical services)1st Floor Plot No. A1/B, IDA Nacharam Cross Road Hyderabad 500 076 Tel: 040-27177036 Fax: 040-27174037 crm@firstsourcels.com; sudhakar@firstsourcels.com;	-do-
4.	Geo Chem Laboratories Pvt. Ltd. Pragati, Adjacent to Crompton Greaves Kanjur Marg (E) Mumbai 400 042 Tel: 022-61915100 Fax: 022-61915101 neel@geochemgroup.com; sureshbabu.p@geochem.net.in;	-do-
5.	Interfield Laboratories XIII/1208, Interprint House Kochi 682 005 Tel: 0484-2217865, 2210915, 221838 mail@interfieldlaboratories.com;	-do-
6.	MicroChem Silliker Pvt. Ltd. MicroChem House A-513 TTC Industrial Area MIDC Mahape Navi Mumbai 400 701 Tel: 022-27787800 deepa@microchem.co.in; dhanya.dhumal@microchem.co.in; vidhya.gangar@microchem.co.in; ajit@microchem.co.in;	-do-
7.	National Collateral Management Services Limited (NCMSL) Team Towers, 4th Floor, Plot No. A-1/2/A Industrial Park IDA-Uppal Hyderabad 500 039 Tel: 040-66374700, 09959333267 Ganesh.r@ncmsl.com; vidya.k@ncmsl.com;	-do-
8.	SGS India Pvt. Ltd. Opposite to State Bank of India 28 B/1 (SP), 28 B/2 (SP) 2nd Main Road Ambattur Industrial Estate Chennai-600 058 Tel: 044-66693109 Fax: 044-24963075 Av.Abraham@sgs.com; manojit.pal@sgs.com;	-do-

9. TUV India Pvt. Ltd. (TUV Nord Pune) Survey No: 423/1 & 3/2 -do-
Near Pashankar Auto (Baner) Sus-Pashan Road Pune 411 021
Tel: 020-67900000 vkgupta@tuv-nord.com; foodlab@tuv-nord.com;
mumbai@tuv-nord.com;
10. TUV Sud South Asia Pvt. Ltd. No. 151, 2nd C Main, 2nd stage -do-
Peenya Industrial Estate Bangalore 560058 Tel: 080-67458000
Fax: 080-67458058 suresh.kumar@tuv-sud.in;
meena.mariappan@tuv-sud.in; farhan.ayesha@tuv-sud.in;
11. National Horticultural Research & Development Foundation (NHRDF) -do-
Pesticide Residue Analysis Laboratory Research Complex
Chittegoan Phata P.O. Darna Sangvi Tq. Niphad Nashik Aurangabad Road
Nashik 422 003 Tel: 02550-237551, 237816 Fax: 237947
hrdf_nsk@sancharnet.in;drpkgupta11@gmail.com;
12. Reliable Analytical Laboratories Pvt. Ltd. 125/139 Indian Corporation -do-
Mankoli Gundavli Bhiwandi Thane 421 302Tel: 02522-398100
harshal@reliablelabs.org; rashmi@reliablelabs.org;
vikas@reliablelabs.org; meenal@reliablelabs.org;
13. Chennai Mettex Lab Private Limited -do-
Jothi Complex No. 83 MKN Road Guindy Chennai 600 032
Tel: 044-22323163, 42179490/91 Fax: 044 22311034
vks@mettexlab.com; drbala@mettexlab.com

Annexure - 7

Method of sampling for pomegranates from the farm/plot to be followed by authorized laboratories/NRL

A. Procedure for Sampling Table Pomegranates for Analysis

1. **Who will draw the sample?**

   Individuals authorized by the nominated laboratory of the APEDA (as per Annexure 6) will only draw the samples.

   - Individuals authorized for the sampling table pomegranates should have letter of authorization from the recognized nominated laboratories.
   - Individuals authorized for the sampling should also have Identity card issued by the laboratory.

2. **From which orchard sample is to be collected?**

   The samples will be drawn only from those orchards which are registered for export with the District Superintending Agriculture / Horticulture Officer of the respective State Govt. Before sampling, following documents pertaining to the registered orchard will be verified/copies obtained by the authorized representative of the laboratory:

   - Registration Certificate issued by the State Government.
   - Registration Record of Pomegranate orchard/ plot (Annexure – 2) and drawing/map lay out.
   - In case the plot drawing/map lay out provided by the Agri/Hort Officer is not fully clear, the laboratory representative may continue to draw the sample as per guidelines given in Section 5 of the main document. However, while doing so, he shall provide clarity to the drawing and obtain the farmer's endorsement on it and provide a copy to him or his representative at the site.
   - Field Pesticide Application Record maintained by the farmer/exporter (Annexure – 2).
   - Second and Final Inspection Report of the Agri/Hort Officer [Annexure –4 (B)]. It is recommended that Annexure – 4 (A) should also be seen.
   - Sample slip signed by the farmer and exporter (Annexure –8).

3. **Locate the block from where the sample is to be drawn?**

- Information given in, Registration Certificate, map lay out and Annexure–2 are to be used to locate the block. It is also expected that the sketch of the block/plot is available with the First Inspection Report [Annexure–4 (A)] of the Agriculture / Horticulture Officer. Thus, the plot may be identified on the basis of name of the block / plot, direction, nearness to the landmarks such as road, well, pump house, shed *etc.*
- Area of the block/plot from where the sample is to drawn should not exceed 4 ha. In case, the area is above 4 ha, additional samples for every 4 ha are to be drawn.
- Area / block / section / plot selected for sampling should have the same expected date of harvest and schedule of pesticide applications. In case, the above aspects are not same, separate sample should be drawn from each variation block. Area should be considered as one section, which should not be larger than 4 ha for collecting one sample.
- Separate Annexure–2, Annexure–4(A), 4(B) and Annexure –8 should be obtained for each section / area / block /plot selected for sampling.
- Area can be determined on the basis of example given below:

  Suppose the row-to-row distance is 15 feet and plant-to-plant distance is 10 feet. Total area occupied by one plant = 150 sq. ft.

  Area of one plot (4 hectares) E" 40, 000 sqm, which is equal to 4, 43, 556 sq. ft Total No. of plants in one plot (4 hectares) = 4, 43, 556/150 = 2, 960 Plants (Approx.)

  So, one plot (Four hectares) contains 2,960 Plants spaced at 15ft. by 10ft. distance.

4. **Collection of Sample**

Sample collected should be most representative to the section/block/area/ plot selected. To ensure the same

- Smallest unit for sampling should be a one fruit, not the part of fruit.
- Size of the fruit selected for sampling should be from the export grade fruits available in block. Sampling of the undersized and oversized fruits to be avoided.

- Fruits hidden in the canopy, or showing infestation of insect pest (thrips, mealy bugs *etc.*) or diseases (Bacterial Blight, spots *etc.*) or any disorder (sun scald, cracking *etc.*) are to be avoided for sampling
- Sample fruits should be collected from all over the section selected for this purpose.

  The selected Section / area / block / plot may be divided into number of spots as shown the diagram below and about 2 plants may be selected from each spot for sampling.

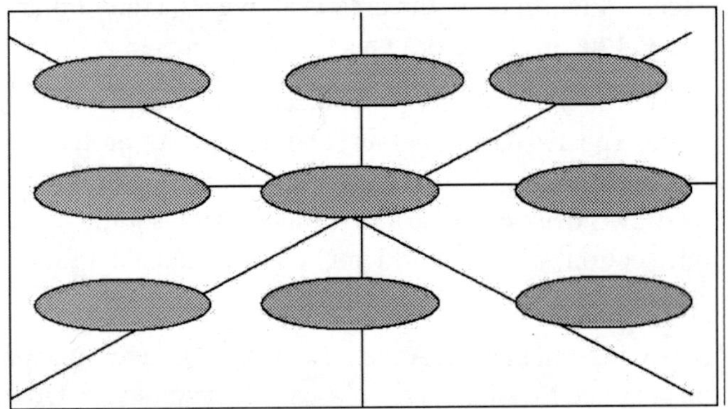

- One sample to be drawn from group of plots having similar package of practices instead of restricting & making it mandatory to draw one sample per plot....Refer Global GAP version IFA4.0_Mar11_AF_CB_FV_4.0-2 clause no CB 8.6
- In any case spots should not be less than 9.
- Selected spot may have about 25 to 30 plants. From these plants, two or three plants may be selected randomly for sampling. However, selected plants should not be abnormal in terms of less canopy, less number of fruits, infected with insect pests, physically damaged *etc.*
- From each selected plant two to three fruits should be harvested for sampling. Fruits should be cut, above the knot on the peduncle with the help of secateurs. Total weight of the sample collected at each spot should be about 1.00 kg (4 to 5 fruits) and will be called as Primary Sample.
- All primary samples from one section will be mixed and will be called as Bulk Sample. The size of the bulk sample should be of 35-40 fruits having minimum 9 kg weight from one plot.

- Laboratory Sample will be drawn by randomly selecting the fruits from the bulk sample. Randomly select fruits for Laboratory Sample (4 kg) (Minimum fruit 20 Fruits) following reduction principle. Out of 4 Kg. samples for laboratory, Pack half of fruits separately in two boxes as Sample for analysis (10 fruits) and for storage (10 Fruits) for storage. Storage fruits are kept in order to revalidate the analysis data in case of any dispute.
- The counter sample should be stored immediately in cold storage at 0°C (or as per guidelines of NRL) with 90-95% relative humidity for a period of 90 days from the date of issue of test report of the sample.
- Data logger should be installed in a cold room for recording temperature and humidity from time to time.
- It is the responsibility of the laboratory to see that the seal of the storage sample is kept intact till such time the sample is required to be in case of dispute.

5. Packing and transport of sample

Two samples should be packed separately in clean and virgin corrugated cardboard box designed for transport of pomegranates. The boxes should be sealed with cellophane tape. Sample slip (Annexure–8) should be kept in polyethylene cover and it should be kept inside the box. The boxes should be labelled from outside with the following information:

- Pomegranate Sample for Residue Analysis
- Sample slip number
- Date of sampling
- Name of authorized representative of the laboratory

Sealed sample should be delivered to the laboratory within 24 hrs of sampling.

6. List of Material required for sampling

- Secateurs or harvesting scissor
- Paper bags or polythene bags
- Paper, markers, pencils

- Hand glove
- Corrugated cardboard boxes of 5 kg capacity
- Labels
- Thread
- Cellophane tape, packing tape
- Seal
- Weighing balance

B. Procedure for sampling soil for monitoring banned agrochemicals

In case of orchards, were the fruits containing banned agrochemicals were found in the last season soil samples shall be collected by following standard procedure and analysed for these chemicals.

C. Procedure for sampling irrigation water for monitoring banned agrochemicals

In case of orchards, were the fruits containing banned agrochemicals were found in the last season water samples shall be collected by following standard procedure and analysed for these chemicals.

Annexure–8

Sample Slip for Pomegranates
(To be given by farmers/lessee farmer/lessee farming organisation / exporters)

No.	FIRST SAMPLE/RE-SAMPLE	Sample Slip No._
(see para 6.9; strike out whichever is not applicable)		
1)	Name & address of the farmer	
2)	Plot Registration No. & validity	
3)	Address/location of the sampled plot	
4)	Crop and variety	
5)	Total area of the plot(s) covered by this sample slip	
6)	Likely production (in MT) declared by Agri/Hort Officer as per Annexure 5(B) covered by this sample slip	
7)	Crop condition pertaining to pests and diseases	
8)	Number of fruits in the sample drawn (per plot)	
9)	Number of fruits in the laboratory sample (including storage sample)	
10)	Date of drawl of sample in the field	
11)	Whether soil or water has been tested (please attach copy of report, if applicable)0	
12)	Pack-house Regn. No. & its validity (if applicable)	

Copy of Annexure – 3 is enclosed.

Date: Signature of Exporter Signature of Farmer/lessee farmer/ lessee farming organization (Name of Exporter) (Name of Farmer)

1. I have drawn this sample personally from the above plot by adopting the procedure given in Annexure–7 of the Procedure for export of pomegranates.
2. This sample is taken from the above plot, which is intended to be exported by (name of the farmer/exporter) and an endorsement to this effect has been made on the plot registration certificate.

3. I have also obtained a copy of the Registration Certificate, Annexure-2 and original of Annexure – 4(B) from the farmer.

4. That, as on date, APEDA recognition of this laboratory is valid.

Date: Signature

Place : Name of authorised Representative of

Nominated Laboratory
Official address:

Annexure – 9

List of agrochemicals to be analyzed for export of Pomegranates

Sr. No.	Name of Chemicals/ Pesticides detected Quantification	Residue Content (mg/kg) Individual	Sum	Harmonized EU-MRL (mg/kg)	Equipment used for analysis	Limit of (LOQ) (mg/kg)
1.	1-Naphthylacetic acid (alphanapthyl acetic acid)	BLQ	BLQ	0.05*	LC-MS/MS	0.05
2.	2,4-D (sum of 2,4-D and its esters expressed as 2,4-D)	BLQ	BLQ	0.05*	LC-MS/MS	0.01
3.	2-Bromo-2-nitropropane- 1,3-diol	BLQ	BLQ	0.01*	GC-MS/MS	0.01
4.	4-bromo-2-chlorophenol (metabolite of Profenophos)	BLQ	BLQ	0.01*	GC-MS/MS	0.01
5.	4- CPA (4 Chlorophenoxy acetic acid)	BLQ	BLQ	0.01*	LC-MS/MS	0.01
6.	6-Benzyl adenine	BLQ	BLQ	0.01*	LC-MS/MS	0.01
7.	Abamectin (sum of avermectin B1a, avermectin B1b and delta- 8,9 isomer of avermectin B1a)	BLQ	BLQ	0.01*	LC-MS/MS	0.01
8.	Acephate	BLQ	BLQ	0.01*	LC-MS/MS	0.01
9.	Acetamiprid	BLQ	BLQ	0.01*	LC-MS/MS	0.01
10.	Alachlor	BLQ	BLQ	0.01*	LC-MS/MS	0.01
11.	Aldrin (Aldrin and dieldrin combined expressed as dieldrin)	BLQ	0.01*	GC-MS/MS	0.01	
11.1	Aldrin	BLQ	BLQ	0.01*	GC-MS/MS	
11.2	Dieldrin	BLQ		0.01*	GC-MS/MS	
12.	Allethrin and Bioallethrin	BLQ	BLQ	0.01*	GC-MS/MS	0.01
13.	Ametoctradin	BLQ	BLQ	0.01*	LC-MS/MS	0.01
14.	Atrazine	BLQ	BLQ	0.05*	LC-MS/MS	0.01
15.	Azadirachtin	BLQ	BLQ	0.01*	LC-MS/MS	0.01
16.	Azoxystrobin	BLQ	BLQ	0.05*	LC-MS/MS	0.01
17.	Benalaxyl including other mixtures of	BLQ	BLQ	0.05*	LC-MS/MS	0.01

	constituent isomers including Benalaxyl-M (sum of isomers)					
18.	Bendiocarb	BLQ	BLQ	0.01	GC-MS/MS	0.01
19.	Benfuracarb	BLQ	BLQ	0.02*	LC-MS/MS	0.01
20.	Benomyl (see carbendazim)	BLQ	BLQ	0.1*	LC-MS/MS	0.01
21.	Bifenazate	BLQ	BLQ	0.01*	LC-MS/MS	0.01
22.	Bifenthrin	BLQ	BLQ	0.05*	GC-MS/MS	0.01
23.	Bitertanol	BLQ	BLQ	0.01*	LC-MS/MS	0.01
24.	Buprofezin	BLQ	BLQ	0.05*	LC-MS/MS	0.01
25.	Butachlor	BLQ	BLQ	0.01*	LC-MS/MS	0.01
26.	Cadmium	BLQ	BLQ	0.05#	ICP	0.02
27.	Captafol	BLQ	BLQ	0.02*	GC-MS/MS	0.01
28.	Captan	BLQ	BLQ	0.02*	GC-MS/MS	0.01
29.	Carbaryl	BLQ	BLQ	0.01*	LC-MS/MS	0.01
30.	Carbendazim (including Benomyl)	BLQ	0.1*	LC-MS/ MS	0.01	
30.1	Benomyl	BLQ	BLQ	0.1*	LC-MS/MS	0.01
30.2	Carbendazim	BLQ		0.1*	LC-MS/MS	
31.	Carbofuran (sum of	BLQ	0.01*	LC-MS/MS	0.01	
31.1	Carbofuran	BLQ	BLQ	0.01*	LC-MS/MS	0.01
31.2	3-hydroxy-carbofuran	BLQ		0.01*	LC-MS/MS	
32.	Carbosulfan	BLQ	BLQ	0.01*	LC-MS/MS	0.01
33.	Carboxin	BLQ	BLQ	0.05*	LC-MS/MS	0.01
34.	Cartap hydrochloride	BLQ	BLQ	0.01*	LC-MS/MS	0.01
35.	Chlorantraniliprole	BLQ	BLQ	0.4	LC-MS/MS	0.01
36.	Chlordane (cis& trans)	BLQ	0.01*	GC-MS/MS	0.01	
36.1	cis-chlordane	BLQ	BLQ	0.01*	GC-MS/MS	0.01
36.2	trans-chlordane	BLQ		0.01*	GC-MS/MS	
37.	Chlorfenapyr	BLQ	BLQ	0.01*	GC-MS/MS	0.01
38.	Chlorfenvinphos	BLQ	BLQ	0.01*	GC-MS/MS	0.01
39.	Chlormequat (CCC)	BLQ	BLQ	0.05*	LC-MS/MS	0.01
40.	Chlorothalonil	BLQ	BLQ	0.01*	GC-MS/MS	0.01
41.	Chlorpyrifos	BLQ	BLQ	0.05*	GC-MS/MS	0.01
42.	Chlorpyrifos methyl	BLQ	BLQ	0.05*	GC-MS/MS	0.01
43.	Clothianidin	BLQ	BLQ	0.02*	LC-MS/MS	0.01
44.	Cyantraniliprole	BLQ	BLQ	0.01*	LC-MS/MS	0.01
45.	Cyazofamid	BLQ	BLQ	0.01*	LC-MS/MS	0.01
46.	Cyfluthrin (including	BLQ	0.02*	GC-MS/MS	0.01	

	other mixtures of constituent isomers sum of isomers)					
46.1	Cyfluthrin 1	BLQ	BLQ	0.02*	GC-MS/MS	0.01
46.2	Cyfluthrin 2	BLQ		0.02*	GC-MS/MS	
46.3	Cyfluthrin 3	BLQ		0.02*	GC-MS/MS	
46.4	Cyfluthrin 4	BLQ		0.02*	GC-MS/MS	
47.	Cymoxanil	BLQ	BLQ	0.05*	LC-MS/MS	0.01
48.	Cypermethrin (including) other mixtures of constituent isomers sum of isomers	BLQ	0.05*	GC-MS/MS	0.01	
48.1	Cypermethrin 1	BLQ	BLQ	0.05*	GC-MS/MS	0.01
48.2	Cypermethrin 2	BLQ		0.05*	GC-MS/MS	
48.3	Cypermethrin 3	BLQ		0.05*	GC-MS/MS	
48.4	Cypermethrin 4	BLQ		0.05*	GC-MS/MS	
49.	Dazomet (Methylisothiocyanate resulting from the use of Dazomet and metam)	BLQ	BLQ	0.02*	LC-MS/MS	0.01
50.	DDT (all isomers, sum of p,p′-DDT, o,p′-DDT, p,p′-DDE and p,p′ -TDE (DDD) expressed as DDT)		BLQ	0.05*	GC-MS/MS	0.01
50.1	p,p′-DDT	BLQ	BLQ	0.05*	GC-MS/MS	0.01
50.2	o,p′-DDT	BLQ		0.05*	GC-MS/MS	
50.3	p,p′-DDE	BLQ		0.05*	GC-MS/MS	
50.4	p,p′-TDE (DDD)	BLQ		0.05*	GC-MS/MS	
51.	Deltamethrin	BLQ	BLQ	0.05*	GC-MS/MS	0.01
52.	Diafenthiuron	BLQ	BLQ	0.01*	LC-MS/MS	0.01
53.	Diazinon	BLQ	BLQ	0.01*	LC-MS/MS	0.01
54.	Dichlorvos	BLQ	BLQ	0.01*	LC-MS/MS	0.01
55.	Dicofol (sum of p, p′ and o, p′ isomers)	BLQ	BLQ	0.02*	GC-MS/MS	0.01
56.	Dieldrin (see Aldrin)	BLQ	BLQ	0.01*	GC-MS/MS	0.01
57.	Difenoconazole	BLQ	BLQ	0.1	LC-MS/MS	0.01
58.	Diflubenzuron	BLQ	BLQ	0.05*	LC-MS/MS	0.01
59.	Dimethoate (Including Omethoate)	BLQ	0.02*	LC-MS/MS	0.01	
59.1	Dimethoate	BLQ	BLQ	0.02*	LC-MS/MS	0.01
59.2	Omethoate	BLQ		0.02*	LC-MS/MS	

60.	Dimethomorph	BLQ	BLQ	0.01*	LC-MS/MS	0.01
61.	Dinocap (sum of dinocap isomers and their corresponding phenols expressed as dinocap) and Meptyldinocap	BLQ	BLQ	0.02*	LC-MS/MS	0.02
62.	Dinotefuran	BLQ	BLQ	0.01*	LC-MS/MS	0.01
63.	Diquat	BLQ	BLQ	0.05*	LC-MS/MS	0.02
64.	Dithianon	BLQ	BLQ	0.01*	LC-MS/MS	0.01
65.	Dithiocarbamates (Mancozeb, Maneb, Propineb, Metiram, Thiram, Zineb and Ziram collectively estimated as CS2)	BLQ	BLQ	0.05*	GC-MS	0.05
66.	Diuron (Diuron including all components containing 3,4-dichloroaniline moiety expressed as 3,4-dichloroaniline)	BLQ	0.01*	LC-MS/MS	0.01	
66.1	Diuron	BLQ	BLQ	0.01*	LC-MS/MS	0.01
66.2	3,4-dichloroaniline	BLQ		0.01*	LC-MS/MS	
67.	Dodine	BLQ	BLQ	0.05*	LC-MS/MS	0.01
68.	Edifenphos	BLQ	BLQ	0.01*	LC-MS/MS	0.01
69.	Emamectin Benzoate	BLQ	BLQ	0.01*	LC-MS/MS	0.01
70.	Endosulphan (All isomers, sum of *alpha*- and *beta*-isomers and endosulphansulphate expressed as endosulphan)	BLQ	0.05*	GC-MS/MS	0.01	
70.1	alpha-Endosulphan	BLQ	BLQ	0.05*	GC-MS/MS	0.01
70.2	beta-Endosulphan	BLQ		0.05*	GC-MS/MS	
70.3	Endosulphan sulphate	BLQ		0.05*	GC-MS/MS	
71.	Endrin	BLQ	BLQ	0.01*	GC-MS/MS	0.01
72.	Ethephon	BLQ	BLQ	0.05*	LC-MS/MS	0.05
73.	Ethion	BLQ	BLQ	0.01*	LC-MS/MS	0.01
74.	Ethofenprox (Etofenprox)	BLQ	BLQ	1	GC-MS/MS	0.01
75.	Etrimfos	BLQ	BLQ	0.01*	LC-MS/MS	0.01
76.	Famoxadone	BLQ	BLQ	0.01*	LC-MS/MS	0.01
77.	Fenamidone	BLQ	BLQ	0.02*	LC-MS/MS	0.01

78.	Fenarimol	BLQ	BLQ	0.02*	LC-MS/MS	0.01
79.	Fenazaquin	BLQ	BLQ	0.01*	LC-MS/MS	0.01
80.	Fenitrothion	BLQ	BLQ	0.01*	GC-MS/MS	0.01
81.	Fenobucarb	BLQ	BLQ	0.01*	LC-MS/MS	0.01
82.	Fenpropathrin	BLQ	BLQ	0.01*	GC-MS/MS	0.01
83.	Fenpyroximate	BLQ	BLQ	0.05*	LC-MS/MS	0.01
84.	Fenthion (fenthion and its oxygen analogue, their sulfoxides and sulfone expressed as parent)	BLQ	0.01*	LC-MS/MS	0.01	
84.1	Fenthion	BLQ	BLQ	0.01*	LC-MS/MS	0.01
84.2	Fenthion-sulfone	BLQ		0.01*	LC-MS/MS	
84.3	Fenthion-sulphoxide	BLQ		0.01*	LC-MS/MS	
85.	Fenvalerate (any ratio of constituent isomers (RR, SS, RS & SR) including esfenvalerate) (F) (R)	BLQ	BLQ	0.02*	GC-MS/MS	0.01
86.	Fenvalerate and Esfenvalerate (Sum of RS & SR isomers) (F)	BLQ	BLQ	0.02*	GC-MS/MS	0.01
87.	Fipronil (sum of fipronil + sulfone metabolite (MB46136) expressed as fipronil)	BLQ	0.005*	GC-MS/ MSor LC- MS/ MS	0.005	
87.1	Fipronil	BLQ	BLQ	0.005*	GC-MS/ MSor LC- MS/MS	0.005
87.2	Fipronil sulfone	BLQ		0.005*	GC-MS/ MSor LC- MS/MS	
88.	Flonicamid (sum of flonicamid, TNFG and TNFA) (R)	BLQ	BLQ	0.05*	LC-MS/MS	0.01
88.1.	Flonicamid	BLQ		0.05*		
88.2	TNFG	BLQ		0.05*		
88.3	TNFA	BLQ		0.05*		
89.	Flubendiamide	BLQ	BLQ	0.01*	LC-MS/MS	0.01
90.	Flufenacet (sum of all compounds containing the N fluorophenyl-N-isopropyl moiety	BLQ	BLQ	0.05*	LC-MS/MS	0.01

	expressed as flufenacet equivalent)					
91.	Flufenoxuron	BLQ	BLQ	0.05*	LC-MS/MS	0.01
92.	Flufenzine	BLQ	BLQ	0.02*	LC-MS/MS	0.01
93.	Fluopicolide	BLQ	BLQ	0.01*	LC-MS/MS	0.01
94.	Flusilazole	BLQ	BLQ	0.01*	LC-MS/MS	0.01
95.	Forchlorfenuron (CPPU)	BLQ	BLQ	0.01*	LC-MS/MS	0.01
96.	Fosetyl-Al (sum fosetyl + phosphorous acid and their salts, expressed as fosetyl)	BLQ	BLQ	75	LC-MS/MS	0.1
97.	Gibberellic acid	BLQ	BLQ	0.01*	LC-MS/MS	0.01
98.	Glufosinate-ammonium (sum of glufosinate, its salts, MPP and NAG expressed as glufosinate equivalents)	BLQ	0.1*	LC-MS/MS	0.05	
98.1	Glufosinate-ammonium	BLQ	BLQ	0.1*	LC-MS/MS	
98.2	MPP	BLQ		0.1*	LC-MS/MS	
98.3	NAG	BLQ		0.1*	LC-MS/MS	
99.	Glyphosate	BLQ	BLQ	0.1*	LC-MS/MS	0.05
100.	HCH (sum of isomers, except the *gamma* isomer)	BLQ	0.01*	GC-MS/MS	0.01	0.01
100.1	alpha-HCH	BLQ	BLQ	0.01*	GC-MS/MS	0.01
100.2	beta-HCH	BLQ		0.01*	GC-MS/MS	
100.3	delta-HCH	BLQ		0.01*	GC-MS/MS	
101.	Heptachlor (sum of heptachlor and heptachlor epoxide expressed as heptachlor)	BLQ	0.01*	GC-MS/MS	0.01	
101.1	Heptachlor	BLQ	BLQ	0.01*	GC-MS/MS	0.01
101.2	Heptachlor epoxide	BLQ		0.01*	GC-MS/MS	
102.	Hexaconazole	BLQ	BLQ	0.01*	LC-MS/MS	0.01
103.	Hexythiazox	BLQ	BLQ	0.5	LC-MS/MS	0.01
104.	Homobrassinolide	BLQ	BLQ	0.01*	LC-MS/MS	0.01
105.	Imidacloprid	BLQ	BLQ	1	LC-MS/MS	0.01
106.	Indoxacarb (sum of R and S isomers)	BLQ	BLQ	0.02*	LC-MS/MS	0.01
107.	Iodosulfuron-methyl (iodosulfuron-methyl including salts, expressed as iodosulfuron-methyl)	BLQ	BLQ	0.01*	LC-MS/MS	0.01

108.	Iprobenphos	BLQ	BLQ	0.01*	LC-MS/MS	0.01
109.	Iprodione	BLQ	BLQ	0.02*	GC-MS/MS	0.01
110.	Iprovalicarb	BLQ	BLQ	0.01*	LC-MS/MS	0.01
111.	Isoprothiolane	BLQ	BLQ	0.01*	LC-MS/MS	0.01
112.	Isoproturon	BLQ	BLQ	0.01*	LC-MS/MS	0.01
113.	Kresoxim methyl	BLQ	BLQ	0.05*	LC-MS/MS	0.01
114.	Lambda-cyhalothrin	BLQ	BLQ	0.02*	GC-MS/MS	0.01
115.	Lead	BLQ	BLQ	0.10#	ICP	0.05
116.	Lindane (*gamma*-HCH)	BLQ	BLQ	0.01*	GC-MS/MS	0.01
117.	Linuron	BLQ	BLQ	0.05*	LC-MS/MS	0.02
118.	Lufenuron	BLQ	BLQ	0.02*	LC-MS/MS	0.02
119.	Malathion (sum of malathion and malaoxon expressed as malathion)	BLQ	BLQ	0.02*	LC-MS/MS	0.01
119.1	Malathion	BLQ	BLQ	0.02*	LC-MS/MS	0.01
119.2	Malaoxon	BLQ	BLQ	0.02*	LC-MS/MS	0.01
120.	Mandipropamid	BLQ	BLQ	0.01*	LC-MS/MS	0.01
121.	Mepiquat	BLQ	BLQ	0.05*	LC-MS/MS	0.05
122.	Metalaxyl & Metalaxyl-M	BLQ	BLQ	0.05*	LC-MS/MS	0.01
123.	Methamidophos	BLQ	BLQ	0.01*	LC-MS/MS	0.01
124.	Methomyl and Thiodicarb (sum of methomyl and thiodicarb expressed as methomyl)	BLQ	BLQ	0.02*	LC-MS/MS	0.01
124.1	Methomyl	BLQ	BLQ	0.02*	LC-MS/MS	0.01
124.2	Thiodicarb	BLQ	BLQ	0.02*	LC-MS/MS	0.01
125.	Metolachlor and S-metolachlor (metolachlor including other mixtures of constituent isomers including S-metolachlor (sum of isomers))	BLQ	BLQ	0.05*	LC-MS/MS	0.01
126.	Metribuzin	BLQ	BLQ	0.10*	LC-MS/MS	0.01
127.	Milbemectin (sum of milbemycin A4 and milbemycin A3, expressed as milbemectin)	BLQ	BLQ	0.02*	LC-MS/MS	0.02
127.1	Milbemycin A3	BLQ	BLQ	0.02*	LC-MS/MS	
127.2	Milbemycin A4	BLQ	BLQ	0.02*	LC-MS/MS	
128.	Monocrotophos	BLQ	BLQ	0.01*	LC-MS/MS	0.01
129.	Myclobutanil	BLQ	BLQ	0.02*	LC-MS/MS	0.01

130.	Novaluron	BLQ	BLQ	0.01*	LC-MS/MS	0.01
131.	Omethoate (refer to Dimethoate)	BLQ	BLQ	0.02*	LC-MS/MS	0.01
132.	Oxadiazon	BLQ	BLQ	0.05*	LC-MS/MS	0.01
133.	Oxycarboxin	BLQ	BLQ	0.01*	LC-MS/MS	0.01
134.	Oxydemeton-methyl (sum of oxydemeton methyl and demeton-S-methylsulfone expressed as oxydemeton methyl)	BLQ	BLQ	0.01*	LC-MS/MS	0.01
134.1	Oxydemeton- methyl	BLQ	BLQ	0.01*	LC-MS/MS	0.01
134.2	Demeton-S-methylsulfone	BLQ	BLQ	0.01*	LC-MS/MS	0.01
135.	Oxyfluorfen	BLQ	BLQ	0.05*	GC-MS/MS	0.01
136.	Paclobutrazol	BLQ	BLQ	0.5	LC-MS/MS	0.01
137.	Paraquat	BLQ	BLQ	0.02*	LC-MS/MS	0.02
138.	Parathion methyl (sum of Parathion methyl and paraoxon methyl expressed as Parathion methyl)	BLQ	BLQ	0.01*	GC-MS/MS	0.01
138.1	Parathion methyl	BLQ	BLQ	0.01*	GC-MS/MS	
138.2	Paraoxon methyl	BLQ	BLQ	0.01*	GC-MS/MS	0.01
139.	Parathion ethyl	BLQ	BLQ	0.05*	GC-MS/MS	0.01
140.	Penconazole	BLQ	BLQ	0.05*	LC-MS/MS	0.01
141.	Pencycuron	BLQ	BLQ	0.05*	LC-MS/MS	0.01
142.	Pendimethalin	BLQ	BLQ	0.05*	LC-MS/MS	0.01
143.	Permethrin (sum of isomers)	BLQ	BLQ	0.05*	GC-MS/MS	0.01
143.1	cis-Permethrin	BLQ	0.05*	GC-MS/MS		
143.2	trans-Permethrin	BLQ	BLQ	0.05*	GC-MS/MS	0.01
144.	Phenthoate	BLQ		0.01*	LC-MS/MS	0.01
145.	Phorate (sum of phorate, its oxygen analogue and their sulfones expressed as phorate)	BLQ	BLQ	0.01*	GC-MS/MSor LC- MS/MS	0.01
145.1	Phorate	BLQ	BLQ	0.01*	GC-MS/MSor LC- MS/MS	
145.2	Phorate-sulfone	BLQ	BLQ	0.01*		0.01
145.3	Phorate-sulfoxide	BLQ		0.01*		
146.	Phosalone	BLQ	BLQ	0.01*	LC-MS/MS	0.01

147.	Phosphamidon	BLQ	BLQ	0.01*	LC-MS/MS	0.01
148.	Pirimiphos-methyl	BLQ	BLQ	0.05*	LC-MS/MS	0.01
149.	Profenophos	BLQ	BLQ	0.01*	LC-MS/MS	0.01
150.	Propamocarb (sum of propamocarb and its salt expressed as propamocarb)	BLQ	BLQ	0.01*	LC-MS/MS	0.01
151.	Propanil	BLQ	BLQ	0.01*	GC-MS/MS	0.01
152.	Propargite	BLQ	BLQ	0.01*	LC-MS/MS	0.01
153.	Propetamphos	BLQ	BLQ	0.01*	GC-MS/MS	0.01
154.	Propiconazole	BLQ	BLQ	0.05*	LC-MS/MS	0.01
155.	Propoxur	BLQ	BLQ	0.05*	LC-MS/MS	0.01
156.	Pyraclostrobin	BLQ	BLQ	0.02*	LC-MS/MS	0.01
157.	Pyridaben	BLQ	BLQ	0.5	LC-MS/MS	0.01
158.	Pyriproxyfen	BLQ	BLQ	0.05*	GC-MS/MS	0.01
159.	Quinalphos	BLQ	BLQ	0.05*	LC-MS/MS	0.01
160.	Simazine	BLQ	BLQ	0.01*	LC-MS/MS	0.01
161.	Spinosad (sum of Spinosyn A+D)	BLQ	BLQ	0.02*	LC-MS/MS	0.01
161.1	Spinosyn A	BLQ		0.02*	LC-MS/MS	
161.2	Spinosyn D	BLQ		0.02*	LC-MS/MS	
162.	Spirodiclofen	BLQ	BLQ	0.02*	LC-MS/MS	0.01
163.	Spiromesifen	BLQ	BLQ	0.02*	LC-MS/MS	0.01
164.	Streptomycin	BLQ	BLQ	0.01*	LC-MS/MS	0.01
165.	*tau*- Fluvalinate	BLQ	BLQ	0.01*	GC-MS/MS	0.01
166.	Tebuconazole	BLQ	BLQ	0.02*	LC-MS/MS	0.01
167.	Temephos	BLQ	BLQ	0.01*	LC-MS/MS	0.01
168.	Tetraconazole	BLQ	BLQ	0.02*	LC-MS/MS	0.01
169.	Tetracycline	BLQ	BLQ	0.01*	LC-MS/MS	0.01
170.	Thiacloprid	BLQ	BLQ	0.02*	LC-MS/MS	0.01
171.	Thiamethoxam (sum of thiamethoxam and clothianidin expressed as thiamethoxam)	BLQ	BLQ	0.05*	LC-MS/MS	0.01
172.	Thiobencarb	BLQ	BLQ	0.01*	LC-MS/MS	0.01
173.	Thiodicarb (see Methomyl)	BLQ	BLQ	0.02*	LC-MS/MS	0.01
174.	Thiometon	BLQ	BLQ	0.01*	LC-MS/MS	0.01
175.	Thiophanate-methyl	BLQ	BLQ	0.10*	LC-MS/MS	0.01
176.	Transfluthrin	BLQ	BLQ	0.01*	GC-MS/MS	0.01

177.	Triadimefon (sum of triadimefon and triadimenol)	BLQ	BLQ	0.1*	LC-MS/MS	0.01
177.1	Triadimefon	BLQ	BLQ	0.1*	LC-MS/MS	
177.2	Triadimenol	BLQ	BLQ			
178.	Triazophos	BLQ	BLQ	0.01*	LC-MS/MS	0.01
179.	Trichlorfon	BLQ	BLQ	0.01*	LC-MS/MS	0.01
180.	Tricyclazole	BLQ	BLQ	0.05*	LC-MS/MS	0.01
181.	Tridemorph	BLQ	BLQ	0.01*	LC-MS/MS	0.01
182.	Trifloxystrobin	BLQ	BLQ	0.02*	LC-MS/MS	0.01
183.	Trifluralin	BLQ	BLQ	0.01*	GC-MS/MS	0.01
184.	Uracil	BLQ	BLQ	1.00†	LC-MS/MS	1

Source : Procedure for Export of Pomegranates, Agricultural and Processed Food Products Export Development Authority, New Delhi

* EU-MRL set at LOQ (mg/kg) as per http://ec.europa.eu/sanco_pesticides/public/index.cfm?event=substance.selection

† These are natural products. EU-MRL does not exist for these chemicals. Hence, their MRL is set at the LOQ of the method developed and validated at the National Referral Laboratory of the NRC for Grapes.

\# Reference : Commission Regulation (EC) No 1881/2006 of 19th December 2006.

(F) = Fat soluble

(R) = Residue definition includes metabolites/isomers

Annexure-10

Electronic Document Certificate of Residue Analysis
(To be issued by the authorized laboratories)

1) Name and address of the farmer
2) Name and address of the exporter
3) Farm/Plot Registration No.
4) Location of the farm/plot
5) Area of the farm/plot(s) covered by this report
6) Total likely production of the farm/plot(s) [in MT] covered by this report [calculated on the basis of Annexure-3 (for area purposes) and Annexure-4 (B)]
7) Name of crop and variety
8) Sample details
   (a) Place and date of sample drawn
   (b) Quantity of sample
   (c) Packing
   (d) Sample code No.
9) Name of the representative of Authorized Laboratory who has drawn the sample
10) Date of drawl of sample
11) Date of receipt of sample in laboratory
12) Date of completion of analysis
13) Packhouse Registration Number & its validity (if applicable)

Sr. No.	Names of chemicals	EU MRL (mg/kg)	Residue content (mg/kg)	Limit of Determination (LOD) (mg/kg)	Method of analysis	Equipment used for analysis
1.	2.	3	4.	5.	6.	7

Certificate

1) This is to certify that the sample was drawn by our authorized representative from farm having Registration No and has been analysed by us. The sample was tested for the residue of the chemicals mentioned above and the residue content in the sample is as given in Column 4 of the table given above.

2) The APEDA recognition of this laboratory is valid as on date.

Result: Sample conforms/does not conform to MRL requirements with respect to the above listed chemicals (strike out whichever is not applicable).

Date : Signature of authorized signatory of Authorized Laboratory alongwith seal

Place :

Annexure - 11

Instructions for grant of C.A. and Certificate of AGMARK grading for exports of pomegranates

Persons desirous of obtaining AGMARK Certification on fruits and vegetables under Agmark should have valid Certificate of Authorization (C.A.) for grading of fruits and vegetables. Provisions contained in Fruits and Vegetables Grading and Marking Rules, 2004 shall be applicable.

I. Procedure for grant of C.A.

1. Persons desirous of obtaining C.A. for grading fruits and vegetables under Agmark for exports shall apply to the concerned office of Directorate of Marketing & Inspection (DMI) in the prescribed Proforma - I.

2. Necessary documents as prescribed in Annex-A shall be enclosed with the application.

3. Demand draft for Rs. 1000/- as C.A. processing fee shall be enclosed with the application.

4. Applicant for grant of C.A. can have his own premises (owned by him or rented). He can also use common facilities of APMC pack houses, Private/Coop pack houses etc. Minimum requirements in the premises are given in Annex-B. Details of such arrangements shall be given with the application. Details of such arrangements may not be given by APEDA approved pack houses.

5. Concerned office of the DMI will process the documents, inspect the proposed premises and grant C.A. within ten days of the receipt of complete documents. Inspection of the premises is not required in case of APEDA approved pack houses. In such cases, CA shall be issued within three days of the receipt of the complete documents.

6. Grade designation mark (AGMARK insignia) shall be securely affixed to or printed on each container. Since each and every container is accounted for in exports, it is not necessary to have running replica

serial No. on each container. It is also not necessary that AGMARK insignia shall be printed in printing presses permitted by the Directorate. However, authorised packer shall inform the name and address of the printing press from whom he is getting the containers bearing AGMARK replica printed.

II. Procedure for obtaining Certificate of AGMARK Grading (CAG) for export of pomegranates to EU countries.

1. C.A. holder shall apply giving details of the consignment to any one of approved laboratories under intimation to the concerned office of DMI for grant of CAG for the lot of pomegranates in the prescribed proforma (Annexure-C). The lists of the offices of DMI and the approved laboratories are at Appendix (i) and Appendix (ii), respectively.

2. The C.A. holder will send the Demand Draft towards grading charges to the laboratory payable @ 0.1% of FOB value subject to a minimum of Rs. 200/- per consignment. The FOB value has been fixed at Rs. 93.20 per kg. The laboratory will send the grading charges to the concerned office of DMI every fortnight. Failure to do so will block the software for the concerned laboratory after a warning.

3. The C.A. holder will offer the lot for inspection at the approved premises. The consignment shall be offered packed in appropriate packing boxes. The inspection may also be carried out on the grading and sorting line of the approved premises.

4. C.A. holder can offer the lot for inspection and grading at the Airport/ Seaport. The size of such lot shall not be more than 5 MT net weight.

5. Approved Chemist of the approved laboratory shall draw sample as per the sampling plan (Annexure – D). He/she will sign on the containers selected for sampling.

6. The approved chemist will grade the sample according to prescribed standards and assign appropriate grade. He/she will fill up the Inspection Report in the prescribed proforma (Annexure – E)

7. The Inspecting Officer will stack-seal the consignment after inspection in the cold store. The temperature of the pomegranates in the cold store should be in the range of 5 – 100 °C and Relative Humidity in the cold store should be in the of range of 90-95%.

8. The Inspecting Officers of DMI can make surprise checks of the grading done by the approved laboratories. They will fill up the

Inspection Reports of such surprise checks. The decision of the Inspecting Officers of DMI shall be final. In case of any dispute, the C.A. holder can refer the matter to the Dispute Settlement Committee.

9. Designated persons of the approved laboratory will issue the CAG in the prescribed proforma. The CAG will be sent electronically to the C.A. holder, concerned office of DMI and the PSC issuing Authority.

10. The CAG shall be valid for 15 days from the date of issue. Revalidation of the CAG can be done on the request of the C.A. holder in case shipment is delayed beyond 15 days for valid reasons. It will be done after re-examination by the concerned laboratory to ascertain that the consignment is in sound merchantable condition and that there has been no deterioration in the quality.

Proforma – I

Application for Grant of Certificate of Authorisation for Grading and Marking of__________(Name of Commodity) for Export Grading

To,

The Dy. Agri. Marketing Adviser/

Asstt. Agri. Marketing Adviser/

Senior Marketing Officer

Directorate of Marketing & Inspection

__________ (name of city)

Sir/Madam,

I/We ________________ of M/s ________________ (full postal address) being desirous of marking [Name(s) of commodity] with a grade designation mark in accordance with the rules made under the provisions of Agricultural Produce (Grading & Marking) Act 1937, hereby, request for grant of Certificate of Authorisation.

I/We have carefully gone through the provisions of AP (G&M) Act, 1937, the General Grading & Marking Rules 1988, relevent commodity Grading & Marking Rules and the instructions issued by the Agricultural Marketing Adviser to the Govt. of India or an Officer authorised by him in this regard for grading & marking of the said commodity and agree to abide by them.

The requisite particulars are furnished herewith in the prescribed proforma and the requisite documents are enclosed.

Yours faithfully,

(Signature of the applicant)

Place : Designation :

Date : for M/s : ______________

ACKNOWLEDGEMENT SLIP

Received the application dated __________ of M/s__________alongwith the enclosures and D.D. No.__________dated __________ for Rs. __________ for grant of Certificate of Authorisation for Grading & Marking of__________ under AGMARK for export grading.

(Office Seal with Signature)

Particulars to be Furnished with the Application for Certificate of Authorisation

1. Name and full postal address of the party.
2. Name(s) of the commodity proposed to be graded.
3. Status of the firm, i.e., Proprietary/Partnership/

   Pvt. Ltd./Public Ltd./Regd. Society/Public

   Undertaking etc. (copy of the relevant document be enclosed).
4. Period for which the applicant has been in the business.
5. Name(s) and address of two representatives of the firm who will attend the grading work and correspond in the matter (specimen signatures to be furnished separately).
6. * (a) RBI Code No., if any

   * (b) Import Export Code No. (issued by DGFT)

   (c) Membership of the Commodity Boards (APEDA, etc.), if any
7. ST/CST No., if allotted.
8. Full address of the premises where grading and marking will be carried out.
9. Status of the said premises owner/lessee
   (strike out whichever is not applicable).

*10. Details of the machinery/packing machines/cold storage etc. available in the plant/premises with their capacity.
Name of the Machinery___Nos. Capacity

11. Any other information relevant to grading of the commodity.
12. Trade name, if any.

(Signature of the applicant/
authorized person) Designation
for M/s

Place :
Date :

* Not required in case of APEDA recognized pack houses.

Annexure - A

List of the documents to be furnished along with the application for grant of C.A. for export grading

1. Application for grant of C.A in the prescribed Proforma-I.
2. Signatures of authorized persons of the firm on the letter pad.

*3. Copy of the proprietorship declaration/partnership deed/ memorandum and articles of association/bye-laws of society etc.

*4. Blue print or neatly drawn sketch of the premises showing all dimensions duly singed by the authorized person of the firm.

5. Medical fitness certificates issued by the Registered Medical Practitioner certifying that the workers engaged in the handling of the product in various operations, are free from any communicable and contagious diseases.

*6. Copy of import export code No. issued by DGFT.

7. Copy of APEDA registration, if registered.

Note : (i) Photocopies of all documents should be signed and stamped by authorized person of the firm.

(ii) Three sets of the documents are to be submitted to the concerned office of the Directorate.

* Not applicable in case of APEDA recognized pack houses.

Annexure - B

Minimum requirements in the premises for grading of fruits & vegetables

1. Premises should be clean and in hygienic condition.
2. Surroundings of the premises should be clean.
3. It should not be situated near tanneries, chemical plants, fertilizer plants etc.
4. Walls of the premises should be properly plastered and free from crevices, holes, dampness etc. Thatched roof is not advisable.
5. Premises should be pest, insect and rodent proof.
6. Premises should be free from cobwebs and spiders.
7. Premises should have proper drainage system.
8. Premises should have facilities for testing of TSS, Sugar-Acid ratio, *etc.* The typical needs of chemicals, apparatus, *etc.*, are given in Appendix – (iii).
9. Premises should have arrangements for disposal of rejected, rotten, waste of horticulture produce.

Annexure – C

To,

(Name of the approved laboratory)

Subject : Request for grant of Certificate of Agmark Grading (CAG) for consignment of pomegranates for export.

Sir,

1. I/We hold Certificate of Authorization No.______valid up to______for grading and marking of fruit and vegetables for exports.
2 I/We intend to export pomegranates to______(destination). Details of the consignment are as follows :
   a) Laboratory test details for pesticide residues.
      Name of the Laboratory Farm Registration No. Test Report No.
   b) Packaging details.

Commodity	No. of Boxes Qty. (in each box)	Total Qty. (in MTs)	FOB value (in Rs.)

3. I/We intend to get the inspection and grading done through your approved laboratory. The above mentioned consignment may be inspected at
   (a) our approved premises at ____________________ .
OR (b) airport/seaport at ____________________ .
4. Demand Draft for Rs.____________________towards grading charges is sent separately.
(5) I/We, propose to effect export of pomegranates referred to above to____________________(destination) and these have been processed and packed under my supervision in the pack house referred to in item (1) above.
(6) I/We, further certify that the pomegranates referred to above are contained in____________________number of boxes/cartons and that the laboratory analysis report establishes that pomegranates do not contain pesticide residues exceeding the MRLs with respect to the destination.
7. It is requested that the CAG may be issued.

Dated : Yours faithfully,

( )

for M/s.

Note: To be e-mailed to the approved laboratory and concerned office of DMI.

Annexure – D

SAMPLING PLAN

No. of cartons in the lot	Minimum No. of cartons to be sampled.
Up to 100	5
101 to 300	7
301 to 500	9
501 to 1000	12
1001 and above	1 % of the cartons (Min 15)

Appendix – (i)

List of offices of the Directorate of Marketing & Inspection

MAHARASHTRA

1. MUMBAI : Dy. A.M.A.
Directorate of Marketing &
Inspection, New CGO, Building
IIIrd Floor, New Marine Lines
Mumbai- 400020.
Telephone No. : 022- 22036801(Direct), 22032699
Fax No. : 22091103
E-mail -dmiromah@nic.in

2. NASIK : Marketing Officer
Directorate of Marketing &
Inspection, New Kamal Niwas,
Behind Hotel Vasco Tourist
Nasik Road - 422101
Telephone No. - 0253-2465437
Fax No. - No fax
E-mail - dmimh05@nic.in

3. SANGLI : Marketing Officer
Directorate of Marketing &
Inspection, APMC Seva Grah Market Yard,
Sangli
Telephone No. - 0233-2670629
Fax No. - No fax
E-mail - dmimh04@nic.in

4. PUNE : Marketing Officer
Directorate of Marketing &
Inspection, Graders Training
Centre, Beej Bhavan, MarketYard, Pune-411007.
Telephone No. - 020-24268598
Fax No. - No fax
E-mail - dmimh07@nic.in

ANDHRA PRADESH

1. HYDERABAD	:	Dy. Agricultural Marketing Adviser Directorate of Marketing & Inspection, Kendriya Sadan Block-1, Sultan Bazar, Hyderabad Telephone No. - 040- 24657446 Fax No. - 040-24731636 E-mail : - dmihyd@ap.nic.in

KARNATAKA

1. BANGALORE	:	Asstt. Agricultural Marketing Adviser Directorate of Marketing & Inspection, APMC Market yard MG Complex, Yashwant Pur, Bangalore-560022 Telephone No. - 080-23472924 Fax No. - 080-23473004 E-mail - bngdmi@kar.nic.in - dmimh05@nic.in

Appendix – (ii)

Laboratories approved by DMI for the grading and marking of fruits and vegetables for export

Sl. No.	Name of the laboratory
01	Pesticide Residue Analysis Laboratory National Horticultural Researches and Development Foundation (NHRDF), P.B. No. 61, Kanada Batata Bhavan, 2954-E New Mumbai Agra Road, Nasik – 422 011.
02	Reliable Analytical Laboratory Pvt. Ltd. Mankoli Naka, Bhiwandi Thane – 421 302.
03	Vimta Labs Ltd., Plot No.5, SP Bio-tech Park, Genome Valley, Shamirpet (M) Hyderabad – 500078.
04	SGS India Ltd., 1/509 A, Old Mahabalipuram Road, Opp. Govt. High School Thoraipakkam, Chennai – 600 085.
05	Shriram Institute for Industrial Research Plot 14 & 15, Sadarmangla Industrial Area White Field Road, Bangalore – 560 048.
06	Interfield Laboratories, XIII/1208A, Interprint House, Kochi – 682005.
07	Delhi Test House, A-62/3, G.T. Karnal Road, Indl. Area,Opp. Hans Cinema, Azadpur Delhi – 110033.
08	M/s ARBRO Pharmaceuticals Ltd., 4/9, Kirti Nagar Industrial Area New Delhi – 110015.
09	M/s.National Collateral Management Services Limited, D.No.4-7-18/6B Raghavendra Nagar, Nacharam, Hyderabad – 500 076.
10	Geo-Chem Lab Pvt. Ltd., 36, Raja Industrial Estate, 1st Floor, Purushottam Kheraj Marg, Mulund (West), Mumbai – 400 080.
11	Pesticide Residue Testing Laboratory Krishibhavan, Shivajinagar, Pune – 411005.
12	Microchem Laboratory Pvt. Ltd.,Microchem House, A-513, TTC Industrial Area MIDC, Mahape Navi Mumbai - 400701.
13	M/S TUV India Private Ltd. 814, Demech House, 2nd Floor, Law College Road Pune – 411004.

Appendix – (iii)

Chemicals, apparatus *etc.* required for evaluating Aril colour in pomegranates

1. To obtain juice from pomegranates	(i) Muslin cloth. (ii) Convenient receptacle (iii) Suitable juice press.
2. Determination of the Total Soluble Solids	(i) A calibrated refractometer or A Brix hydrometer of suitable range, calibrated in tenths of a percentage and standardized at 200 °C. (ii) Thermometer of 0° to 50 °C.
3. Determination of acid content	(i) 20 ml pipette (ii) 50 ml burette (iii) 250 ml conical flask (iv) 250 ml beaker (v) Suitable bottles with labels to store sodium hydroxide, sulphuric acid, phenolphthalein, sodium carbonate, distilled water, etc. (vi) Sodium hydroxide, A.R. (250g) for making 0.1333 N solution. (vii) Sulfuric acid - sp. gr. 1.84 - A.R. 98% pure (500 ml) for making 0.1333 N solution. (viii) Phenolphthalein in ethyl alcohol, 0.4% (minimum size pack). (ix) Sodium carbonate 0.1 N (1 litre standard solution). (x) Distilled water.

Annexure 'E'

INSPECTION REPORT FOR POMEGRANATES

1. Name of the commodity : ____________________
2. Name of the authorized packer : ____________________
3. Address of the pack house : ____________________
4. Lot No./Batch No. : ____________ 5. Shipping mark, (if any) : ____________
6. No. of Boxes X Qty. in each box = total quantity.

   ____________ ____________ ____________

   Quality parameters.
7. Cleanliness : ________ 8. Soundness : ________ 9. Foreign matter ________
10. Pests : ____________ 11. General appearance ____________
12. Damage caused by pests or disease : ____________________
13. Abnormal external moisture : ____________________
14. Foreign smell/taste : ____________________
15. Damages caused by high/low temperature : ____________________
16. Visible traces of moulds : ____________________
17. Condition of the Arils : ____________________
18. Aril size (if applicable) : ____________________
19. Total Soluble Solids : ____________________
20. Sugar/acid ratio : ____________________
21. Defects in shape : ____________________
22. Defects in colour : ____________________
23. Defects in skin by sun scorch : ____________________
24. Bruising : ____________________
25. Skin defects : ____________________
26. Size (weight in grams) : ____________________
27. Percentage Grade Tolerances : ____________________
28. Remarks (if any): ____________________
29. Grade assigned : ____________________

Recommended /not Recommended for issue of Certificate of AGMARK Grading.

(Signature)

Name of the approved chemist and the laboratory

Dated : ____________

MINISTRY OF AGRICULTURE

(Department of Agriculture and Co-operation)
New Delhi, the 14th June, 2004)

G.S.R 220. – Whereas the draft of the Fruits and Vegetables Grading and Marking Rules, 2003 were published as required by Section 3 of the Agricultural Produce (Grading and Marking) Act, 1937 (1 of 1937) at pages 2065 – 2132 of the Gazette of India, Part II, Section 3, Sub section (i) dated 20.9.03 vide GSR 335, dated 3rd September, 2003 for inviting objections and suggestions from all persons likely to be affected thereby;

And whereas copies of the said Gazette were made available to the public on 21st September, 2003;

And whereas the objections and suggestions received from the public in respect of the said draft rules have been duly considered by the Central Government;

Now, therefore, in exercise of the powers conferred by Section 3 of the Agricultural Produce (Grading and Marking) Act, 1937 (1 of 1937), and in super session, of (1) the Pomegranates Grading and Marking Rules, 1937, (2) the Plums Grading and Marking Rules, 1938, (3) the Onion Grading and Marking Rules, 1964, (4) the Banana Grading and Marking Rules, 1980, (5) the Mangoes Grading and Marking Rules, 1981, (6) the Pineapple Grading and Marking Rules, 1982, (7) the Guavas Grading and Marking Rules, 1996 and (8) the Garlic Grading and Marking Rules, 2002, except as respects things done or omitted to be done before such supersession the Central Government hereby makes the following rules namely :-

RULES :

1. Short title, application and commencement:-
   - (i) These rules may be called the Fruits and Vegetables Grading and Marking Rules, 2004.
   - (ii) They shall apply to commercial varieties of Fruits and Vegetables.
   - (iii) They shall come into force from the date of their publication in the Official Gazette.

2. Definitions :-

   (i) "Agricultural Marketing Adviser" means the Agricultural Marketing Adviser to the Government of India;

   (ii) "Authorised packer" means a person or a body of persons who has been granted a certificate of authorization to grade and mark Fruits and Vegetables in accordance with the grade standards and procedure prescribed under these rules;

   (iii) "Certificate of Authorisation" means a certificate issued under the provisions of the General Grading and Marking Rules, 1988 authorising a person or a body of persons to grade and mark Fruits and Vegetables with the grade designation mark;

   (iv) "General Grading and Marking Rules" means the General Grading and Marking Rules, 1988 made under section 3 of the Agricultural Produce (Grading and Marking) Act, 1937 (1 of 1937);

   (v) "Grade designation" means a designation prescribed as indicative of the quality of fruits and vegetables;

   (vi) "Grade designation mark" means the "AGMARK Insignia" referred to in Rule 3;

   (vii) "Schedule" means a Schedule appended to these Rules.

3. Grade designation mark - The grade designation mark shall consist of "AGMARK insignia" consisting of a design incorporating the certificate of authorization number, the word "AGMARK", name of commodity and grade designation resembling the design as set out in Schedule – I.

4. Grade Designation and Quality - The grade designation and quality of Fruits and Vegetables shall be as set out in schedules II to XIX.

5. Method of Packing:-

   (i) Fruits and Vegetables shall be packed in such a way as to protect the produce properly.

   (ii) The materials used inside the package must be new, clean and of a quality such as to avoid causing any external or internal damage to the produce.

   (iii) The use of materials particularly of paper or stamps bearing trade specifications is permitted provided the printing or labeling has been done with non toxic ink or glue.

(iv) Fruits and Vegetables shall be packed in each container in compliance with the Recommended International Code of Practice for Packaging and Transport of Tropical Fresh Fruit and Vegetables (CAC/RCP 44-1995) for export and as per the instructions issued by the Agricultural Marketing Adviser from time to time for domestic market.

(v) The containers shall meet the quality, hygiene, ventilation and resistance characteristics to ensure suitable handling, shipping and preserving of the Fruits and Vegetables. Packages must be free of harmful foreign matter and obnoxious smell.

(vi) Content of each package or lot must be uniform and contain only Fruits and Vegetables of same origin, variety and grade designation.

(vii) The visible part of the contents of the package (if present) must be representative of the entire content.

(viii) Contents of package may have different fruits and vegetables of different varieties/grades as per buyer requirements with proper labeling.

6. Method of Marking and Labelling:-

(i) The grade designation mark shall be securely affixed to or printed on each package in a manner approved by the Agricultural Marketing Adviser or an officer authorized by him in this behalf.

(ii) Following particulars shall be clearly and indelibly marked on each package namely :-

(a) Name of the commodity;

(b) Variety;

(c) Grade designation;

(d) Size code (if prescribed);

(e) Lot/batch/code number;

(f) Country of origin;

(g) Net weight/No. of units;

(h) Name and address of the packer/exporter;

(i) Best before date (where applicable);

(j) Storage conditions, if any;

(k) Date of packing;

(l) Such other particulars as may be specified by the Agricultural Marketing Adviser.

(iii) The ink used for marking on packages shall be of such quality which may not contaminate the product.

(iv) The authorized packer may, after obtaining the prior approval of the Agricultural Marketing Adviser, mark his private trade mark or trade brand on the graded packages provided that the same do not indicate quality other than that indicated by the grade designation mark affixed to the graded packages in accordance with these rules.

7 Fruits and Vegetables may be graded and marked as per buyer requirements for exports provided the minimum requirements specified in the relevant Schedule are met.

8. For domestic trade, Fruits and Vegetables shall comply with the residue levels of heavy metals, pesticides, aflatoxin and other food safety parameters as specified in Prevention of Food Adulteration Rules, 1955.

9. Special conditions of certificate of authorization - In addition to the conditions specified under sub-rule (8) of the Rule 3 of the General Grading and Marking Rules, 1988, every authorized packer shall follow all instructions prescribed by Agricultural Marketing Adviser from time to time.

SCHEDULE – I

(See rule 3)

(Design of AGMARK Insignia)

Name of Commodity ______________________

Grade ______________________

Appendix – (ii)

Grade Designation and Quality of Pomegranate

1. Pomegranates shall be fruits obtained from Varieties (cultivars) of plant *Punica granatum* L. of Puniaceae family;
2. Minimum Requirements:
   - (i) Pomegranates shall be:
     - (a) fresh in appearance;
     - (b) mature and solid in feel;
     - (c) clean, free from any visible foreign matter;
     - (d) free from pests affecting the general appearance of the produce;
     - (e) free of damage caused by pests;
     - (f) free of cracking of skin, mechanical injury/rubbing, staining;
     - (g) free of abnormal external moisture excluding condensation following removal from cold storage;
     - (h) free of any foreign smell or taste;
     - (i) free of any pronounced blemishes.
   - (ii) Pomegranates should not be affected by rotting or deterioration such as to make it unfit for consumption.
   - (iii) Pomegranates shall comply with the residue levels of heavy metals, pesticides and other food safety parameters as laid down by the Codex Alimentarius Commission for exports.
3. Criteria for Grade Designatin :

Grade Designation	Grade Requirements	Grade Tolerances
Extra class	Pomegranate in this class must be of superior quality. They must have the shape, development and colouring that are typical of the variety and/or commercial type. They must be free of defects, with the exception of very slight superficial defects,	5% by number or weight of pomegranates not satisfying the requirements of the grade, but meeting those of class

	provided these do not affect the general appearance of the produce, the quality, the keeping quality and presentation in the package.	I grade or, exceptionally, coming within the tolerances of that grade.
Class I	Pomegranates in this class must be of good quality. They must be characteristics of the variety and/or commercial type. The following slight defects may be allowed, provided these do not affect the general appearance of the produce, the quality, the keeping quality and presentation in the package. - a slight defect in shape. - a slight defect in colouring; - slight skin defects (i.e. scratches, scars, scraps and blemishes) not exceeding 5% of the total surface area	10% by number or weight of pomegranates not satisfying the requirements of the class, but meeting those of class II or, exceptionally, coming within the tolerances of that grade.
Class II	This class includes pomegranates which do not qualify for inclusion in higher classes, but satisfy the minimum requirements. Following defects may be there provided the pomegranates retain their essential characteristics as regard the quality, the keeping quality and presentation: - defects in shape; - defects in colouring - skin defects (i.e., scratches, scars, scrapes and blemishes), not exceeding 10% of the total surface area.	10% by number or weight of pomegranates not satisfying the requirements of the grade, but meeting the minimum requirements.

4. Other Requirements

Pomegranates must be carefully picked and have reached an appropriate degree of development and ripeness in accordance with criteria proper to the variety and/or commercial type and to the area in which they are grown. The development and condition of the Pomegranate must be such as to enable them;

- to withstand transport and handling, and
- to arrive in satisfactory condition at the place of destination.

5. Provisions Concerning Sizing

Size is determined by the weight or maximum diameter of the equatorial section of the fruit, in accordance with the following table:

Size code (minimum)	Weight in grams (minimum)	Diameter in mm.
A	400	90
B	350	80
C	300	70
D	250	60
E	200	50

Size tolerance

(i) For all grades, 10% by number or weight of pomegranate corresponding to the size immediately above and/or below than indicated on the package.

(ii) The maximum size range of 8 mm. between fruit in each package is permitted.

Annexure-12

Electronic Document

Specimen format of declaration (To be given by the exporter on their letterhead {scanned copy} to the PSC issuing authority/laboratory)

1) I, resident of, have/operate from packhouse having APEDA Packhouse Recognition No. dated valid up to and which is located at the following address:

2) I/We, hereby, certify that__________MTs of pomegranates have been procured for export from plot(s) bearing plot registration numbers as given below after drawl of samples as per the procedure prescribed in Annexure - 7 of the Procedures for Export of Pomegranates to the issued by APEDA.

   a) __________renewed on__________.

   b) __________renewed on __________.

   c) __________renewed on __________ etc.

3) The laboratory analysis reports bearing No.__________ dated __________ pertains to the pomegranates quantities referred to in para (2) above.

4) I/We propose to effect export of the pomegranates referred to above to __________(destination) and these have been processed and packed under my supervision in the packhouse referred to in para (1) above.

5) I/We certify that the pomegranates referred to above are contained in __________ number of boxes/cartons and that the laboratory analysis report establishes that the pomegranates do not contain chemicals residues exceeding the MRLs with respect to the destination, referred to in para (4) above, stated in Annexure – 9 of the Procedures for Export of Pomegranates.

6) I/We certify that I/we have satisfied my/ourselves that the relevant EU Regulations as on date as regards the product quality and residues of chemicals have been complied with in respect of the pomegranates referred to above.

7) I/We certify that I/we have verified the registration records (as given in the format of Annexure- 2) as well as Annexure - 4(B) of the plot(s) from

where pomegranates have been harvested for this consignment and that the plot(s) fulfil(s) the procedure laid down in the Procedures for Export of Pomegranates.

8) I/We certify that the consignment covered by this declaration does not contain pomegranates from unregistered plot(s) or from plot(s) whose registration has been cancelled/suspended or from plot(s) that have not cleared the residue tests prescribed by the procedure contained in the Procedures for Export of Pomegranates.

9) I/We certify that, as on this date, the NRL has not issued any Internal Alert Information in respect of the samples drawn by them from the pack house (referred to in para - 1 above) and from the farms/plot(s) (referred to in para - 2 above).

OR

It is certified that the NRL had issued an alert for Plot Registration No. _ vide Internal Alert Information No. _ and, subsequently, the same has been revoked vide their Notification No. after re-sampling.(strike out whichever is not applicable)

10) I/We certify that the AGMARK inspection of the above consignment has been carried out by (name of laboratory) and that the CAG No.________ pertains to the above consignment.

11) I/We certify that the above information/declaration is true and correct.

Date: Signature of Authorized Signatory
Place: of Exporter/Farmer Name and address

Annexure-13

Electronic Document

Internal Alert Information

(To be issued by NRL)
Phone: 020-6914245; Fax: 020-6914246;
E-mail: nrcgrape@mah.nic.in

Alert Information	Original

Sub: Detection of _______ agrochemicals beyond MRLs Page:

No of Pages

1. Name of the commodity and variety :
2. Farm/Plot Registration No. :
3. Code Number of the produce, if any :
4. Date of harvest :
5. Date of sampling :
6. Place of sampling : Farm/Plot
   Pack-house
7. Period of analysis : .... to ....
8. Findings of the analysis

   ____

   ____

9. Recommendations by National Referral Laboratory

   ____

   ____

Date :
Place :

Signature of the Coordinator/
Authorized Signatory of NRL
along with seal

1. Concerned Agriculture/Horticulture Officer
2. State Governments
3. All PSC issuing authorities
4. APEDA, New Delhi
5. All authorized laboratories
6. Growers' Federation
7. Farmers' Association
8. Exporters' Association

Source : Procedure for Export of Pomegranates. Agricultural and Processed Food Products Export Development Authority, New Delhi 110 016.

6

Future Prospects

6.1 Strength of India

- In India, pomegranate is cultivated over 1.31 lakh ha with a production of 13.46 lakh tonnes and a productivity of 10.27 t/ha.
- India is the largest producer of pomegranates in the world.
- Pomegranate is currently ranked 10th in terms of fruit consumed annually in the world. India is the only country in the world where pomegranate is available throughout the year i.e. From January to December. It is cultivated in 3 seasons (*Ambia bahar, Mrig bahar and hasth bahar*) in Deccan plateau of India.
- India is endowed with wide agro climatic conditions that offer immense scope for cultivation of various kinds of fruit crops. This provides an excellent platform for the country to emerge as a leading producer of fruit crop.
- India produces finest varieties of pomegranate having soft seeds, very less acids and very attractive colour of the fruits and grains.
- The versatile adaptability, hardy nature, low maintenance cost, steady but high yields, better keeping quality, fine table and therapeutic values and possibilities to throw the plant into rest period when irrigation potential is generally low, indicate the avenues for increasing the area under pomegranate in India.
- There is strong research support for scientific cultivation of pomegranate like National Research Center for Pomegranate, Solapur, MPKV, Rahuri in Maharashtra and IIHR, Bangalore in Karnataka state.
- Pomegranate co-operative societies from Maharashtra state have formed an apex cooperative namely MAHA ANAR.
- Bhagwa variety has high acceptance in European market.

- Pomegranate export facility center is being set up/has been set up in Baramati area with mechanical handling system.
- Farmers have been trained for export quality production and have registered with GLOBALGAP certification.
- MSAMB has recently obtained brand name i.e. "MAHAPOM".

6.2 Developmental Approach

- Introduction of *Punica protopunica*, wild pomegranate from Socotra Island and its conservation for further utilization in breeding programme. This has to be used for screening against bacterial blight disease and may also be used for widening the genetic base of pomegranate by interspecific hybridization .
- Harnessing of natural biodiversity: Survey, collection and conservation of 'field variants/ off-types' of pomegranate with desirable traits (high yield, better quality, tolerant to diseases etc.) from farmer's field.
- Identification of resistant/ tolerant source for bacterial blight disease of pomegranate caused by *Xanthomonas axonopodis* pv. *punicae* through rigorous screening of all indigenous collection (IC), exotic collection (EC), *etc.* using challenge inoculation method and/or suitable molecular techniques.
- Large scale raising of seedling population of the probable source of tolerance (Nana, Daru) to bacterial blight and rigorous screening against bacterial blight.
- Development of hybrids resistant/ tolerant to bacterial blight through hybridization between commercial cultivars and bacterial blight tolerant genotypes available in the field gene banks (FGB).
- Development of wilt resistant varieties of pomegranate and screening the germplasm for wilt tolerance.
- Development of transgenic pomegranate variety resistant to pests (pomegranate fruit borer) and diseases (Bacterial blight, wilt).
- Development of varieties/hybrids free from physiological disorders *viz.*, fruit cracking, internal breakdown, sunscald.
- Development of pomegranate varieties / hybrids suitable for table purpose with enhanced yield and better quality.
- Development of pomegranate varieties/ hybrids suitable processing purpose: Varieties with better juice recovery for juicing purpose; Varieties with soft seeds, high acidity (>3.0%) suitable for *anardana* purpose; varieties with high sugar content and moderate acidity suitable for wine making.

- Identification and involving of pomegranate varieties with highly pliable rind for development of hybrids resistant to cracking.
- Development of pomegranate varieties bearing fruits of multi-coloured arils (so that it is preferred over monocolour arils).
- Varieties for hostile climate: Development of pomegranate varieties withstanding the challenges of climate: tolerant to frost, hailstorm injury; tolerant to flood/ water stagnation.
- Development of pomegranate varieties with only hermaphrodite flowers (without male flowers & intermediate flowers) so that fruit set occurs in all the flowers after pollination / fertilization and ultimately boost the yield.
- Development of pomegranate varieties with less sugar content (suitable for diabetic patients) so that juice and arils are suitable for diabetic patients.; Development of pomegranate varieties rich in carotenes (so that it helps aged people for vision); Development of pomegranate varieties with more affinity for specific nutrients and that results in fruits rich in specific nutrients: calcium rich fruits for lactating mother; iron, iodine rich fruits for pregnant women.
- Value added products from pomegranate: Pomegranate flour for use as baby food (similar to banana flour); Pomegranate tonic for growing children; pomegranate candies/fruit bars; pomegranate chewing gums; tablets for preparing instant pomegranate tea; with high carotenes for aged people with blurred vision.
- Development of ornamental pomegranate varieties: Double flower, Nana type, bonsai types etc suitable for indoor gardening, hanging baskets, *etc.*
- Development of short, medium duration varieties of pomegranate by hybridizing with 'precocious' early types so that ROI is quicker.
- Special pomegranate varieties: Development of extra-dwarf varieties suitable for 'meadow orchard'; Development of varieties suitable for protected cultivation in shadenet house, polyhouse,; Development of varieties suitable for hydroponics technology.
- Development of new varieties/ hybrids with resistance to biotic and abiotic stresses using Marker Assisted Selection, Marker Assisted Back Cross, Marker Assisted Recurrent Selection, gene pyramiding, Genome wide selection for interested traits for crop improvement; Gene silencing technique to address the problem of bacterial blight, fruit borer, *etc.*; Development of complete linkage map for chormosomes of pomegranate; Understanding the population structure and evolutionary relationship of accessions.

- Use of genomics (structural, functional & comparative) for basic, applied & strategic research.
- Development of Doubled Haploidy (DH) lines in pomegranate for reduction of breeding cycle and release of varieties as per the need.
- Improving water use efficiency (WUE) of pomegranate during different phonological stages and seasons.
- Enhancing nutrient use efficiency (NUE) of pomegranate for sustainable production.
- Enhanced production strategy based on INM and increased NUE, WUE for sustainable production.
- Establishing mother nurseries for production and supply of quality planting material.
- Expansion of pomegranate to potential non-traditional areas: Kullu and other apple growing areas in HP, other areas of western Himalaya would be covered with pomegranate due to climate change.
- Use of renewable energy sources for production and processing of pomegranate.
- Mechanized farm operations for pesticides spray, pruning and harvesting of pomegranate.
- Community farming in pomegranate: Due to high cost, pomegranates are affordable only to high class people; by 2050, the fruits would be affordable and consumed by all (including below poverty line people) due to reduction in cost/ price; thus, domestic consumption will be more; Integrated farming system approach including animal husbandry, poultry, fisheries and pomegranate.
- Intensification of organic production thereby soil organic carbon will increase upto a level in equilibrium with the existing environmental condition.
- The large scale pomegranate processing industry will come up in the country along with fragmented small scale processing units in production catchment areas.
- Potential of pomegranate will be utilized to the fullest extent for nutraceutical purpose and medicines from pomegranate for heart diseases, cancer, oral health, skin health, reproductive health will be commonly available and become multi billion industry.
- All the processing units will utilize pomegranate completely making it highly profitable venture and thereby reducing the cost of processed products.

Appendix-A

List of Pomegranate Exporters from India

RCMC No.	Exporter Name	Address	City	State	PIN	FAX	Mobile	Telephone	E-mail
239	ASAR BROTHERS	124-A, ADARSH INDL. ESTATE, SAHAR ROAD, ANDHERI (E),	MUMBAI	Maharashtra	400099	28370612	9820232426	67681324	asarbrothers@gmail.com
309	ASHRAF EXPORTS	279, NAGDEVI STREET, 1 ST FLOOR	MUMBAI	Maharashtra	400003	23441705	9820084891	23439977	ashrafexports@vsnl.net
560	ASHVINA TRADING CO	9, HASTAGIRI, ASHOK CHAKRAVARTHY ROAD, KANDIVALI (EAST).	MUMBAI	Maharashtra	400101	28872636	9820046578	28876658	ashvina1@ashvinatrading.com
153134	BAGBAN TRADING COMPANY	(PROP. NASIR MOHD. SHAFI BAGWAN) P.O. BOX NO. 48, A/P. TANANG TASGAON PHATA, PANDHARPUR RD., NEAR ACC CEMENT GODOWN	MIRAJ	Maharashtra	416416	2223279	9422041911	2231396	nsbagban@yahoo.com
164546	BARAKAH EXIM,	1003, QUEENSGATE BLDG, HIRANANDANI- ESTATE, G.B.ROAD	THANE WEST	Maharashtra	400607	27846364	9967764750	27846364	azzi.shaikh@gmail.com
1393	BHAGWANDAS BHERUMAL & CO	26/27, SITARAM BUILDING, OPP: CRAW FORD MARKET,	MUMBAI	Maharashtra	400001	23428104	9820041915	23438955	bbcexpo@bom3.vsnl.net.in
2265	BOMBAY FRUITS &	33/35, MATRU VATSALYA GROUND FLOOR,ROOM NO.2,	MUMBAI	Maharashtra	400002	22406075	9821018946	22408642	bombayexp@vsnl.com

	VEGETABLES IMPORT EXPORT PVT. LTD.	2ND FOFALWADI, BHULESHWAR,							
154772	C.K. EXPORT IMPORT	GS-33,GR.FLR,CENTRAL FACILITY BLDG., APMC FRUIT MARKET,VASHI	MUMBAI	Maharashtra	400705	27848494	9867722202	27842090	ckexportimport@gmail.com
51013	CHAND FRUIT COMPANY PRIVATE LIMITED	GAT NO.154/ 2A+2B,TAKLI ROAD, MIRAJ, TALUKA MIRAJ	SANGLI	Maharashtra	416410	2223279	9422041911	2228532	shafibag@yahoo.com
157724	DEEPAK FERTILISERS AND PETRO CHEMICALS CORPORATION LIMITED	OPP.GOLF COURSE, SHASTRI NAGAR, YERAWADA,	PUNE	Maharashtra	411006	66478080	9975689009	66478006	devendra.gupta@dfpcl.com
154033	DIMPLE TRADING COMPANY	DHEERAJ HERITAGE, OFFICE NO.20, 4TH FLOOR, S.V. ROAD, NEAR MILAN SUBWAY, SANTACRUZ WEST	MUMBAI	Maharashtra	400054	26601865	9323486780	26601565	dimpletc@gmail.com
162173	DINESH BALKRISHNA JAGDALE	SHINDE WASTI, AJANALE, TAL. SANGOLA	SOLAPUR	Maharashtra	413308	27830112	8097771678	9975831570	diamond_dinesh@yahoo.co.in
164660	ELEGANT EXPORTS	LAKE PALACE, 4B-306, POWAI VIHAR COMPLEX	POWAI, MUMBAI	Maharashtra	400076	25703215	25703215	elegant.exports@hotmail.com	

51185	EVEREST GLOBAL	ROYAL PALMS, 319 MASTER MIND 1 AAREY MILK COLONY, GOREGAON (E)	MUMBAI	Maharashtra	400065	28792323	9821175743	28794400	everestg@vsnl.com
720	FOODS AND INNS LTD	FOODS & INNS BUILDING, SION-TROMBAY ROAD, PUNJABIWADI, DEONAR	MUMBAI	Maharashtra	400088	23533106	9820397988	23533104	fni@vsnl.com
156160	GLOBAL RESOURCES	304, SHEETAL APTS., NARAYAN NAGAR,SION CHUNNABHATTI (E),	MUMBAI	Maharashtra	400022	24051199	9820054547	24052211	arshall.machado@gmail.com
3062	GNT EXPORTS	J-510-511, APMC FRUIT MARKET, SECTOR-19, TURBHE, VASHI	NAVI MUMBAI	Maharashtra	400705	27841440	9820415049	27841749	info@gntexports.com
158039	GREEN VEGE EXPORTS PVT. LTD.	B-48/01, Chintamani Hsg. Society, Sector 26, Vashi	NAVI MUMBAI	Maharashtra	400703	41236014	9819213951	27882417	greenvege_exportpvtltd@rediffmail.com
	INI FARMS PVT. LTD.	B-202, UNIVERSAL BUSINESS PARK, NEAR KAMANI OILS, SAKIVIHAR RD, CHANDIVALI, ANDHERI (E),	MUMBAI	Maharashtra	400072	42600700	8108178842	42600700	info@inifarms.com
2920	J. P. CORPORATION	B-84/3, MULUND COLONY, NEAR SHANKAR PALACE HOTEL, L.B.S. MARG, MULUND (W)	MUMBAI	Maharashtra	400082	25671751	9820221532	25671750	jvpingle@gmail.com
5649	KALYA EXPORTS	A-2, AJANTA COMPLEX, NEW ADGON NAKA, PANCHAVATI	NASHIK	Maharashtra	422003	2513004	9423176953	2513001	kalyaexports@yahoo.co.in
1649	KAY	606, DEV CORPORA, OPP.	THANE	Maharashtra	400602	41578901	9870565640	41578900	export@kaybeeexports.com

	BEE EXPORTS	CADBURY COMPANY, HIGHWAY, KHOPAT	WEST						
152980	KONKAN AGRO	27, SAHAR CARGO ESTATE, V. M. SHAH ROAD, V. M. SHAH ROAD, J. B. NAGAR, ANDHERI (EAST),	MUMBAI	Maharashtra	400099	28261458	9820632318	28261458	samarth3@mtnl.net.in
152498	KRUTIKA EXPORTS	PLOT NO.R-310, ROAD NO.10, Maharashtra MIDC, RABALE,	NAVI MUMBAI	400701	27603792		9820092069	27691672	kraftpack2007@mtnl.net.in
155291	KSHIRSAGAR COLD STORAGE	PROP. MRS. BHARTI M. KSHIRSAGAR) AT SHIVDI, POST UGAON, TAL. NIPHAD,	NASHIK	Maharashtra	422304	2550244207	9822327671	244207	kshirsagarcoldstorage@ rediffmail.com
158138	LAKSHMI IMPEX	801, Kailash Tower, Shiv Shrishti Complex, Goregaon Link Road, Mulund (W)	MUMBAI	Maharashtra	400080	25686300	9969810322	25606300	lakshmiexpo@rediffmail.com
155756	LELE AGRO EXPORTS	NO.15, SHRINIWAS BLDG., L.B.S. MARG, OPP. KURLA COURT, KURLA - WEST	MUMBAI	Maharashtra	400070	25430389	9028016885	25340389	leleagro@gmail.com
	M AND M ENTERPRISES	303, KARISHMA, MAHANT ROAD EXT.	VILE PARLE (EAST), MUMBAI	Maharashtra	400057	26105708	9820199892	66964301	mandmenterprises2 @gmail.com
163640	MADHAV GLOBAL TRADE	22, MADHAV, AMRUT VARSHA COL.,SAINATH NAGAR,	NASHIK	Maharashtra	422006	2597882	9823086907	2597882	madhavglobaltrade @gmail.com
154564	MAHA ANAR	C/O. AKHIL MAHARASHTRA DALIMB UTPADA SANSHODHAN SANGH PUNE, E-15, NISARG,	PUNE	Maharashtra	411037	24266620	9922986664	24266620	mahaanar@yahoo.co.in

		GULTEKADI MARKET YARD							
168711	NATIONAL IMPEX	SHOP NO. 13, SUCCESS CHAMBERS BEHIND NAGAR VACHNALAYA RAJWADA	SATARA	Maharashtra	415001	282191	9623775399	mkrasai@gmail.com	
150103	NEERAJ INTER-NATIONAL	117/119, KAZI SYED STREET	MUMBAI	Maharashtra	400003	24161848	9820308611	23446174	anand_neeraj@hotmail.com
157217	P.C. FOODS PVT. LTD.	04, NIKITA APARTMENT, 65, DATEY NAGAR, NEAR ATHARVA MANGAL KARYALAY, GANGAPUR ROAD	NASHIK	Maharashtra	422005	2342898	9923388595	2342896	info@pcfoods.co.in
50860	PANACEA ENERGIZERS PVT LTD	A/202, Shreenath Complex A, P. K. Extn Road, Nahur, Mulund (W),	MUMBAI	Maharashtra	400080	25917375	9820339406	25624509	paresh@panaceaexim.com
5530	R.S. ENTERPRISES	B-2, DEVADIGA C.H.S., A WING, FLAT NO. 1, OM NAGAR,. J. B. NAGAR, ANDHERI (E)	MUMBAI	Maharashtra	400099	28390569	9820632318	28355838	samarth3@mtnl.net.in
159549	RAINBOW INTER-NATIONAL (PROP. ABHIJEET C. BHASALE)	7, SAMARTH PRASAD, 122, TULSHIBASWALE COLONY, SAHKARNAGAR,	PUNE	Maharashtra	411009	24213093	9890093363	24213093	abhi@rainbowinternational.in
	RELIABLE FRESH	S. NO. 33/13, FLAT NO. A-2, RAVI TERRACE, BEHIND BHARTI VIDHYAPEETH,	PUNE	Maharashtra	411046	24377088	9822553208	24360273	reliablefresh@yahoo.com

		AMBEGAON ROAD, AMBEGAON							
157166	RGS AGRO IMPEX	B/1/05, 1ST FL, GROMA HOUSE, PLOT NO.14C, SECTOR -19, VASHI	NAVI MUMBAI	Maharashtra	400705	27891713	9821163048	27891713	rgsagroimpex@hotmail.com
165466	SADGURU SUPPLIERS	NEW SATARA SOC, B 103, PLOT NO 2/3, SECTOR 14, KOPARKHAIRNE,	NAVI MUMBAI	Maharashtra	400701	27881293	9892227294	27881293	ashoksarjine@yahoo.in
154480	SANGOLA AGRO PVT.LTD	201, 165-D, RAILWAY LINES, VARDA APARTMENTS, SOLAPUR	SOLAPUR	Maharashtra	413001	260285	9423066638	260285	sheetalchandane @rediffmail.com
166571	SUNITA EXPORTS	B/1601, VRINDAVAN TOWER, PADMA NGR, CHIKUWADI, BORIVALI- WEST,	MUMBAI	Maharashtra	400092	28334943	9004757453	28334943	info@sunita-exports.com
157558	USHA INTER-NATIONAL	134, MODY STREET, 2ND FLOOR,R. NO. 5, SAWANT CHAMBERS, FORT	MUMBAI	Maharashtra	400001	22692265	9167283888	22690690	ushainternationalexpo @gmail.com
5460	VIJAYSHREE EXPORTS	GATE NO. 87, STATION ROAD, AT. SHIVADI, POST. UGAON, TAL. NIPHAD	NASIK	Maharashtra	422304	244307	9822327671	244207	vijayshreeexports@ rediffmail.com.

Source: http://www.nrcpomegranate.org/List_of_Pomegranate_Exporters_From_India.pdf

Appendix-B

List of Pesticides / Pesticides Formulations Banned in India

A.	Pesticides Banned for manufacture, import and use (28 Nos.)
1.	Aldrin
2.	Benzene Hexachloride
3.	Calcium Cyanide
4.	Chlordane
5.	Copper Acetoarsenite
6.	Cibromochloropropane
7.	Endrin
8.	Ethyl Mercury Chloride
9.	Ethyl Parathion
10.	Heptachlor
11.	Menazone
12.	Nitrofen
13.	Paraquat Dimethyl Sulphate
14.	Pentachloro Nitrobenzene
15.	Pentachlorophenol
16.	Phenyl Mercury Acetate
17.	Sodium Methane Arsonate
18.	Tetradifon
19.	Toxafen
20.	Aldicarb
21.	Chlorobenzilate
22.	Dieldrine
23.	Maleic Hydrazide
24.	Ethylene Dibromide
25.	TCA (Trichloro acetic acid)
26.	Metoxuron
27.	Chlorofenvinphos
28.	Lindane (Banned vide Gazette Notification No S.O. 637(E) Dated 25/03/2011)-Banned for Manufecture,Import or Formulate w.e.f. 25th March,2011 and banned for use w.e.f. 25th March,2013.

B.	Pesticide / Pesticide formulations banned for use but their manufacture is allowed for export (2 Nos.)
1.	Nicotin Sulfate
2.	Captafol 80% Powder

C.	Pesticide formulations banned for import, manufacture and use (4 Nos)
1.	Methomyl 24% L
2.	Methomyl 12.5% L
3.	Phosphamidon 85% SL
4.	Carbofuron 50% SP
D.	**Pesticide Withdrawn (7 Nos)**
1.	Dalapon
2.	Ferbam
3.	Formothion
4.	Nickel Chloride
5.	Paradichlorobenzene (PDCB)
6.	Simazine
7.	Warfarin

LIST OF PESTICIDES REFUSED REGISTRATION

S. No.	Name of Pesticides
1.	Calcium Arsonate
2.	EPM
3.	Azinphos Methyl
4.	Lead Arsonate
5.	Mevinphos (Phosdrin)
6.	2,4, 5-T
7.	Carbophenothion
8.	Vamidothion
9.	Mephosfolan
10.	Azinphos Ethyl
11.	Binapacryl
12.	Dicrotophos
13.	Thiodemeton / Disulfoton
14.	Fentin Acetate
15.	Fentin Hydroxide
16.	Chinomethionate (Morestan)
17.	Ammonium Sulphamate
18.	Leptophos (Phosvel)

PESTICIDES RESTRICTED FOR USE IN INDIA

S. No.	Name of Pesticides
1.	Aluminium Phosphide
2.	DDT
3.	Lindane
4.	Methyl Bromide
5.	Methyl Parathion
6.	Sodium Cyanide
7.	Methoxy Ethyl Mercuric Chloride (MEMC)
8.	Monocrotophos
9.	Endosulfan*
10.	Fenitrothion
11.	Diazinon
12.	Fenthion
13.	Dazomet

Endosulfan* :- Endosulfan has been banned by the Supreme Court of India w.e.f. 13-05-2011 for production, use & sale, all over India, till further orders vide ad-Interim order in the Writ Petition (Civil) No. 213 of 2011.

Source : Directorate of Plant Protection, Quarantine & Storage Central Insecticides Board & Registration Committee, Faridabad.

Appendix-C

Conversion factors

Area		
1 ha	=	2.47100 acres
	=	100 x 100 sq m
	=	50 Nali
1 Nali	=	200 sq m
1 acre	=	0.40468 ha
	=	4840 sq yd
	=	43560 sq ft
	=	0.00156 sq miles
	=	8 Kanals
1 sq mile	=	640 acres
	=	259 ha
	=	2.59 sq km
1 sq kilometer	=	0.3861 sq miles
100 sq ft	=	9.290 sq m
10 Marlas	=	0.0253 ha
Yield		
100 kg per ha	=	1.4869 bushels (60 lb) per acre
1 bushel (60 lb) per acre	=	67.253 kg per ha
1 kg per ha	=	0.892169 lb per acre
1 lb per acre	=	1.120864 kg per ha
Energy		
1 BTU	=	251.9958 calorie
	=	0.0003930148 horsepower-hour
	=	1055.056 Joule
	=	0.0002930711 kilowatt-hour
	=	0.2930711 watt hour
1 calorie	=	0.003968321 BTU
	=	4.1868 Joule
	=	0.001163 watt-hour
Weight		
1 metric ton	=	2204.7 pounds
	=	0.98421 long ton
	=	1 .10231 short ton

	=	1000 kg
	=	10 q
1 q	=	100 kg
1 long ton	=	2.240 pounds
	=	1.01605 metric tons
1 short ton	=	2.000 pounds
	=	0.90781 metric ton
1 kg	=	2.20462 pounds
1 pound	=	0.45359 kg
1 ounce	=	0.283 kg or 28.35 gm
16 ounce	=	1 pound
10 pounds	=	4.54 kg
1 mound	=	81.2857 pounds
	=	37.32410 kg
Distance		
1 mile	=	8 furlongs or 1760 yards or 5280 ft
	=	1.609344 km
	=	1609.344 m
1 furlong	=	220 yards
	=	660 ft
	=	0.201168 km
	=	201.168 m
	=	0.125 mile
1 km	=	5/8 mile = 0.6213712 mile = 4.97097 furlong
	=	3280.83 ft
	=	1000 m
	=	1093.613 yards
1 meter	=	1.0936 yards
	=	3.28089 ft
	=	100 cm
1 inch	=	25.4 mm
	=	2.54 cm
1 foot	=	30.48 cm
	=	0.3048 m
1 yard	=	91.44 cm
	=	0.9144 m

Atmosphere to lbs per sq inch	14.73
British thermal unit to kilo calorie	0.252
Cubic centimeter to cubic inch	0.061103
Cubic feet to cubic metre	0.02832
Cubic feet to gallons	6.228
Cubic inch to litre	0.01639
Cubic metre to cubic yard	1.308
Foot lbs per second to horse power	0.001818
Foot lbs to kilogram metre	0.1383
U.S. Gallons to litre	3.785
Barrels to gallons	42.00
Grams to ounces	0.03527
Grams to lbs	0.002205
Horse power to kwatts	0.746

Source: Fertiliser Statistics, 2006-07, The Fertiliser Association of India, New Delhi.

References

Adaskaveg, J.E., and H. Forster. 2003. Management of gray mold of pomegranates caused by *Botrytis cinerea* using two reduced-risk fungicides, fludioxonil and fenhexamid. *Phytopathology.* **93:**S127.

Adsule, R. and Patil, N.B. 2005. Pomegranate. **In:** *Handbook of Fruit Sciences and Technology Production, Composition, Storage and Processing,* Salunkhe DK, Kadam SS (Eds). Marcel Dekker, New York, printed at Brijbasi Art Press Ltd, UP, India, pp 455-464.

Artes, F., J.A. Tudela, and R. Villaescusa. 2000. Thermal postharvest treatment for improving pomegranate quality and shelf life. *Postharv. Biol. Rech.* **18:**245–251.

Aulakh, P.S., and H.S. Sur. 1999. Effect of mulching on soil temperature, soil moisture, weed population, growth and yield in pomegranate. *Progr. Hort.* **31:**131–133.

Babu, K.D., Chandra, R., Sharma, J. and Jadhav, V.T. 2009b. Floral biology of pomegranate cultivar Ganesh under Solapur conditions of Maharashtra. *Souvenir and Abstracts of 2nd International Symposium on Pomegranate and Minor Including Mediterranean Fruits,* June 23-27, 2009, University of Agricultural Science, Dharwad, India, pp. 86-87.

Beam Home. 2007. *Punica granatum.* Pomegranate yields showy blooms, then edible fruit. Available online: www.greenbeam.com/features/plant041299.stm

Ben-Arie, R., and E. Or. 1986. The development and control of husk scald on 'Wonderful' pomegranate fruit during storage. *J. Am. Soc. Hort. Sci.* **111:**395.

Blumenfeld, A., F. Shaya, and R. Hillel. 2000. Cultivation of pomegranate. *Options Me´diterrane´ennes Se´rie A, Se´minaires Me´diterrane´ens.* **42:**143–147.

Bretschneider, E. 1935. *History of European Botanical Discoveries in China* (Vols 1, 2). Sampson Low, Marston and Co., London, pp. 1-345.

Bucsbaum, H., D. Baum, and H. Younis. 1982. Weed control in pomegranates. *Phytoparasitica.* **10:**276.

Chadha, K.L.2005. *Handbook of Horticulture*, Directorate of Information and Publication of Agriculture, Indian Council of Agricultural Research, New Delhi, India, pp. 297-304.

Chatterjee, D. and G.S. Randhawa, 1952. Standardized names of cultivated plants in India. Fruits. *Indian J. Hort.*, **9:** 24–36.

Defilippi, B.G., B.D. Whitaker, B.M. Hess-Pierce, and A.A. Kader. 2006. Development and control of scald on Wonderful pomegranate during long-term storage. *Postharv. Biol. Tech.* **41:**234–243.

J., and P. Luddar. 1987. Influence of sodium salts on the Na, Cl and SO_4 contents in leaves, shoots and roots of *Punica granatum* L. *Gartenbauwissenschaft.* **52:**26–31.

Evreinoff, V.A. 1949. Le grenadier. *Fruits Journal of Agriculture Sciences.* **19:** 320-336.

Fahan, A. 1976. The leaf. pp. 171–212. The flower. pp. 321–394. The seed. pp. 419–430. In: Plant anatomy. Hakkibutz Hameuhad Publi., Jerusalem.

Feng, Y., Zheng Chen, D.J., Song, M.T., Zhao, Y.L. and Li, Z.H. 1998. Assessment and utilization of pomegranate varieties resources. *Journal of Fruit Science.* **15 (4):** 370-373.

Fouad, M.M., Barkat, M.R. and El-yazal, S.A. 1979. Bud burst activity, flowering and fruit set of 'Grenouillere' and 'Manfaloti' pomegranate cultivars under Giza conditions. *Research Bulletin of Faculty of Agriculture and Animal Science, Shams University*, pp. 1089-1092.

Ghatge, P.U., D.N., Kulkarni, A.B. Rodge, and Kshirsagar, R.B. 2005. Studies on post-harvest treatments for increasing storage life of pomegranate. *J. Soil. Crop*. **15:**319–322.

Goor, A. and Liberman, J. 1956. *The Pomegranate.* J. Atsmon (ed.), State of Israel, Ministry of Agriculture, Agr. Publ. Section, Tel Aviv, pp. 5–57.

Hays, W.B. 1957. *Fruit growing in India.* 3rd ed., Kitabistan, Allahabad, India, pp. 408–411.

Hepaksoy, S., U. Aksoy, H.Z. Can, and Ui, M.A. 2000. Determination of relationship between fruit cracking and some physiological responses, leaf characteristics and nutritional status of some pomegranate varieties. *Options Me'diterrane'ennes Se'rie A, Se'minaires Me'diterrane'ens*. **42:**87–92.

Hess-Pierce, B.M., and Kader, K.K. 2003. Response of 'Wonderful' pomegranates to controlled atmosphere. *Acta Hort.* **600:**751.

Hodgson, R.W. 1917. *The pomegranate.* Bulletin of California Agricultural Experiment Station, **76:** 163-192.

Holland, D. Hatib, K. and Bar-Ya'akov, I. 2009. Pomegranate: Botany, horticulture, breeding. **In:** *Horticultural Reviews* (Vol 35) Janick J (Ed). John Wiley & Sons, New Jersey, pp 127-191.

http://agritech.tnau.ac.in/horticulture/horti_fruits_pomegranate.html

http://apeda.gov.in/apedawebsite/

http://nhb.gov.in/Default.aspx

http://nhm.nic.in/

http://www.cibrc.nic.in/

http://www.iihr.res.in/

http://www.nrcpomegranate.org/

Hunt, T. 1989. *Plant Names of Medieval England*, DS Brewer, Cambridge, UK, pp 1-390.

Josan, J.S., Jawanda, J.S. and Uppal, D.K. 1979. Studies on the floral biology of [21 cultivars of] pomegranate. I. Sprouting of vegetative buds, flower bud development, flowering habit, time and duration of flowering and floral morphology. *Punjab Horticultural Journal.* **19 (1/2):** 59-65.

Joshi, B.C., 1956. A contribution to the morphology of *Punica granatum* L. *Thesis*, Agra, University, Agra.

Kader, A.A. 2006. Postharvest biology and technology of pomegranates. pp. 211–220. In: N.P. Seeram, R.N. Schulman, and D. Heber (eds.), Pomegranates:ancient rts to modern medicine. CRC Press Taylor & Francis Group, Boca Raton, FL.

Kader, A.A., A. Chordas, and S.M. Elyatem. 1984. Responses of pomegranates to ethylene treatment and storage temperature. *Calif. Agr.* **38(7&8):**14.

Kanwar, Z.S., and D.P. Thakur. 1973. Controlling soft rot of pomegranate (*Punica granatum* L.) fruits caused by *Rhizopus arrhizus* Fischer by using growth regulators and antibiotics. *Haryana J. Hort. Sci.* **2:**50–55.

Karale, A.R., V.S., Supe, S.N. Kaulgud, and P.N. Kale. 1993. Pollination and fruit set studies in pomegranate. *J. Maharashtra Agr. Univ.* **18:**364–366.

Labuda, R., K. Hudec, E. Piekova, J. Mezey, R. Bohovic, J. Mateova, and S.S. Lukac. 2004. *Penicillium implicatum* causes a destructive rot of pomegranate fruits. *Mycopathologia.* **157:**217–223.

Levin, G.M. 1978. The floral biology of pomegranate (*Punica granatum* L.) in south-west Turkmenistan. *Turkmenistan SSR Ylymlar Akademijasynyn Habarlary Biologik Ylymlaryn.***5:**31–38.

Levin, G.M. 2006. Pomegranate roads: a Soviet botanist's exile from Eden. pp. 15–183. B.L.Baer (ed.), Floreat Press, Forestville, CA.

Marathe, R.A., Chandra, R., Jadhav, V.T. and Singh, R. 2009. Soil and nutritional aspects in pomegranate (*Punica granatum* L.). *Environment and Ecology.* **27:** 630-637.

Mars, M. 1996. Pomegranate genetic resources in the Mediterranean region. **In:** *Proceedings of Ist MESFIN Plant Genetic Resources Meeting, Tenerife,* Spain. pp 345-354.

Mars, M. 2000. Pomegranate plant material genetic resources and breeding: a review. *Seminaires Mediterraneens.* **42:** 55-62.

Melgarejo, P., J.J. Martinez, F. Hernandez, F.R. Martinez, P. Barrows, and A. Erez. 2004. Kaolin treatment to reduce pomegranate sunburn. *Sci. Hort.* **100:**349–353.

Mirdehghan, S.H., and M. Rahemi. 2005. Effects of hot water treatment on reducing chilling injury of pomegranate (*Punica granatum*) fruit during storage. *Acta Hort.* **682:**887–892.

Mirdehghan, S.H., M. Rahemi, M. Serrano, F. Guillen, D. Martinez-Romero, and D. Valero. 2005. Prestorage heat treatment to maintain nutritive and functional properties during postharvest cold storage of pomegranate. *J. Agr. Fd Chem.* **54:**8495–8500.

Moghadam, E.G., and S. Nikkhah. 2005. Effect of the postharvest treatment in reducing the pomegranate moth damage. Proc. *Int. Conf. Postharv. Tech. Quality Managem. in Arid Tropics*, Sultanate of Oman, 31 Jan.–2 Feb., 195–198.

Mohamed, A.K.A. 2004. Effect of gibberellic acid (GA3) and benzyladinine (BA) on splitting and quality of Manfalouty fruits. *Assiut J. Agr. Sci.* **35(3):**11–21.

Morton, J. 1987. *Pomegranate*. pp. 352–355. **In:** *Fruits of warm climates.* Miami, FL. Available via DIALOG. https://www.hort.purdue.edu/newcrop/morton/pomegranate.html. Accessed 26 July 2016

Morton, J. 1987. Pomegranate. pp. 352–355. In: Fruits of warm climates. Miami, FL. www.hort.purdue.edu/newcrop/morton/pomegranate.html.

Mukerjee, P.K. 1958. Storage of pomegranates (*Punica granatum* L.). *Sci. Cult.* **24:**94.

Naeini, M.R., A.H. Khoshgoftarmanesh, and E. Fallahi. 2006. Partitioning of chlorine, sodium and potassium and shoot growth of three pomegranate cultivars under different levels of salinity. *J. Plant Nutr.* **29:**1835–1843.

Naeini, M.R., A.H. Khoshgoftarmanesh, H. Lessani, and E. Fallahi. 2004. Effects of sodium chloride induced salinity on mineral nutrients and soluble sugars in three commercial cultivars of pomegranate. *J. Plant Nutr.* **27:**1319–1326.

Nalawadi, U.G., A.A. Farqi, Dasappa, M.A.N. Reddy, Gubbaiah, G.S. Sulikeri, and A.S. Nalini. 1973. Studies on the floral biology of pomegranate (*Punica granatum* L.). *Mysore J. Agr. Sci.* **7:**213–225.

Nath, N. and Randhawa, G.S. 1959a. Classification and description of some varieties of *Punica granatum* L. *Indian Journal of Horticulture.* **16:** 191-200.

Nath, N. and Randhawa, G.S. 1959b. Studies on floral biology of pomegranate. II. Anthesis, dehiscence, pollen studies and receptivity of stigma. *Indian Journal of Horticulture.* **16:** 121-135.

Or-Mizrahi, E., and R. Ben-Arie. 1984. Preventing husk scald in pomegranate cv. 'Wonderful'. *Hasade.* **64:**2464–2468.

Palou, L., C.H. Crisosto, and D. Garner. 2007. Combination of postharvest antifungal chemical treatments and controlled atmosphere storage to control gray mold and improve storability of 'Wonderful' pomegranates. *Postharv. Biol. Tech.* **43:**133–142.

Pareek, O.P. and Sharma, S. 1993. Genetic resources of under exploited fruits. In: Chadha KL, Pareek OP (Eds) *Advances in Horticulture* (Vol 1), Malhotra Publication, New Delhi, pp. 189-225.

Parmar, C. and Kaushal, M.K. (1982). *Wild Fruits of the Sub-Himalayan Region*, Kalyani Publishers, Ludhiana, pp.74-79.

Patil, V.K. and Waghmare, P.R. 1983. Tolerance level of exchangeable sodium percentage (ESP) for pomegranate. *Journal of Maharashtra Agricultural University.* **8:** 257-259.

Porat, R., B. Weiss, Y. Fuchs, A. Neria, A. Gizis, A. Tzviling, D. Gamersny, A. Sharaby-Nov, and R. Ben-Arie. 2005. Long-term storage of 'Wonderful' pomegranate fruit by modified and controlled atmposphere technologies. *Alon Hanotea*. **59:**396 399.

Porat, R., B. Weiss, Y. Fuchs, A. Zandman, and I. Kosto. 2007. Storage of pomegranate fruit in large modified atmosphere bags in harvest bins. *Alon Hanotea*. **61:**714–717.

Prasannakumar, B. 1998. Pomegranate. In: Chattopadhyay TK (Ed) *A Text- book on Pomology* (Vol 3), Kalyani Publishers, Ludhiana, 232 pp.

Pruthi, J. S. and Saxena, A. K. 1984. Studies in anardana (dried pomegranate seeds). *Journal of Fd Science and Technology.* **21:** 296-299.

Ram Asrey, Singh, R.B. and Shukla, H.S. 2002. Effect of sodicity levels on growth and leaf mineral composition of pomegranate (*Punica granatum* L.). *Annals of Agricultural Research.* **23:** 398-401.

Ranjbar, H., A. Sarkhosh, and M. Ghasemnegad. 2006. The effects of calcium chloride, hot water treatment and polyethylene bag packaging on the storage life and quality of pomegranate (cv. Malas-Saveh). p. 113. In: ISHS, 1st Int. *Symp., Pomegranate and Minor Mediterranean Fruits*, Abstracts contributed papers, 16–19 Oct., Adana, Turkey.

Rao, G.G. and Khandelwal, M.K. 2001. Performance of ber (*Zizyphus mauritiana*) and pomegranate (*Punica granatum* L.) on sandy loam saline and saline black soils. *Indian Journal of Soil Conservation.* **29:** 59-64.

Ravid, N., M. Richker, S. Shemesh, R. Birger, F. Abed Elhadi, I. Azmon, and S. Antman. 2004. Soil mulching with plastic sheets in orchards: the idea and the current results. *Alon Hanotea.* **58:** 318–321.

Rodov, V., Z. Schmilovitch, B. Ronen, A. Hoffman, H. Egozi, R. Porat, G. Goldman, B. Horev, B. Weiss, Y. Vinokur, I. Shomer, and D. Holland. 2005. Mechanically separated pomegranate arils: a new lightly processed fresh product. p. 11. In: Fresh-Cut Expo, Phoenix 2005. 5th IFPA & S-294 Poster Session, 14–16 Apr., Phoenix, AZ.

Sachs, Y., G. Ward, T. Agar, and R. Porat. 2006. Preserving pomegranate quality in modified atmosphere packaging. p. 102. In: ISHS, 1st *Int. Symp., Pomegranate and Minor Mediterranean Fruits*, Abstracts contributed papers, 16–19 Oct., Adana, Turkey.

Saeed, W.T. 2005. Pomegranate cultivars as affected by Paclobutrazol, salt stress and change in fingerprints. *Bull. Faculty Agr., Cairo Univ.* **56:**581–615.

Saied, R., S.M. Saneii, and M.A. Asgari. 2003. Selective control of dodder (*Cuscuta monogina* L.) parasitizing pomegranate trees (*Punica granatum* L.). *Acta Hort.* **620:**215–219.

Seeram, N.P., Y. Zhang, J.D. Reed, C.G. Krueger, and J. Vaya. 2006. Pomegranate phytochemicals. pp. 3–29. In: N.P. Seeram, R.N. Schulman, and D. Heber (eds.), *Pomegranates: ancient rts to modern medicine.* CRC Press Taylor & Francis Group, Boca Raton, FL.

Sepahi, A. 1986. GA3 concentration for controlling fruit cracking in pomegranates. *Iran Agr. Res.* **5:**93–99.

Sharma, K.K., Sharma, J. and Kumar, P. 2006. *Important diseases, disorders and insect-pest of pomegranate and their management.* Technical Bulletin of the National Research Centre on Pomegranate. **1:** 1-16.

Sheik, M.K., and M.M. Rao. 2006. Effect of chemicals on control of fruit cracking in pomegranate (*Punica granatum* L.) var. Ganesh. p. 62. In*: ISHS, 1st Int. Symp., Pomegranate and Minor Mediterranean Fruits*, Abstracts contributed papers, 16–19 Oct., Adana, Turkey.

Shmilovich,Z.,Y. Sarig,B. Ronen,A. Hofman,H. Egozi,H. Beres,E. Bar-Lev,andF. Groz. 2006. Development of a method and system for extracting the seeds (arils) from pomegranate fruits—from concept to commercial utilization. p. 127. In:*ISHS, 1stInt. Symp.,Pomegranate and Minor Mediterranean Fruits,* Abstracts contributed papers, 16–19 Oct., Adana, Turkey.

Singh, D.B., B.D. Sarma, and R. Bhargava. 2003. Effect of boron and GA3 to control fruit cracking in pomegranate (*Punica granatum*). *Current Ag.* **27:**125–127.

Singh, R.P., Kar, P.L. and Dhuria, H.S. 1978. Studies on the behaviour of flowering and sex expression in some pomegranate cultivars. *Plant Science.* **10:** 29-31.

Singh, S.S., Krishnamurthi, S. and Katyal, S.L. 1967. *Fruit Culture in India,* Indian Council of Agricultural Research, New Delhi, pp. 207-214.

Tabatabaei, S.Z., and A. Sarkhosh. 2006. Analysis and comparison of salinity tolerance among 10 Iranian commercial pomegranate cultivars. p.54. In: *ISHS 1st Int. Symp. Pomegranate and Minor Mediterranean Fruits*, Abstracts contributed papers, 16–19 Oct, Adana, Turkey.

Tedford, E.C., J.E. Adaskaveg, and A.J. Ott. 2005. Impact of Scholar (a new post-harvest fungicide) on the California pomegranate industry. *Plant Health Progress.* Feb.:1–3.

Tous, J. and Ferguson, L. 1996. Mediterranean fruits. In: *Progress in New Crops,* Janick J (Ed). ASHS Press, Arlington, VA, pp 416-430.

Trapaidze, T.G., and L.S. Abuladze. 1998. Pomegranate cultivars resistant to cracking. *Subtropi cheskie Kul'tury.* **2:**95–97.

Tzulker, R., I. Glazer, I. Bar-Ilan, D. Holland, M. Aviram, and R. Amir. 2007. Antioxidant activity, polyphenol content, and related compounds in different fruit juices and homogenates prepared from 29 different pomegranate Accessions. *J. Agr. Fd Chem.,* http://pubs.acs.org/cgibin/abstract.cgi/jafcau/asap/abs/jf071413n.html.

Vyas, N.L., and K.S. Panwar. 1976. A new post-harvest disease of pomegranate in India. *Current Sci.* **45:**76.

Wang, H.X. 2003. The characteristics of Mudanhua pomegranate variety and its cultural techniques. *South China Fruits.* **32:**49–50.

Yazici, K., Ulger, S. and Kaynak, L. 2009. The relationship between flower development and endogenous gibberellins in pomegranate cv. 'Hicaznar'. *Abstracts of 2nd International Symposium on Pomegranate and Minor Including Mediterranean Fruits*, University of Agricultural Science, Dharwad, India, June 23-27, 2009, 108-109.

Yazici, K., Ulger, S. and Kaynak, L. 2009. The relationship between flower development and endogenous gibberellins in pomegranate cv. 'Hicaznar'. *Abstracts of 2nd International Symposium on Pomegranate and Minor Including Mediterranean Fruits*, University of Agricultural Science, Dharwad, India, June 23-27, 2009, 108-109.

Zukovskij, P.M. 1950. Punica. In: *Cultivated plants and their wild relatives.* State Publ. House Soviet Science, Moscow, pp. 60–61.

Colour Plates

Chapter 2: Production Technology

Fig. 2.1 : Arakta

Fig. 2.2 : Bhagwa

Fig. 2.3 : Mridula

Fig. 2.4 : Ruby

Fig. 2.5 : Dholka

Fig. 2.6 : G-137

Improved varieties of pomegranate

Fig. 2.7 : An attractive fruit of Bhagwa variety

Fig. 2.8 : Arils colour of different varieties of pomegranate

Fig. 2.9 : Air- layering

Fig. 2.10 : Plant tissue culture unit

Fig. 2.11 : Single node segment proliferated into multinodal shoot

Fig. 2.12 : Ideal plant geometry

Fig. 2.13 : Drip irrigation for better yield

1. Bermuda grass

***Coynodon dactylon* (L.) Pers. (Poaceae)**

2. Purple nut sedge *Cyperus rotundus* L. (Cyperaceae)

3. Flat sedge

Cyperus iria (Cyperaceae)

4. Tropical spiderwort *Commelina benghalensis L.* (Commelinaceae)

5. Swine cress *Didymus* (L.) Sm. (Brassicaceae)

6. Annual brachiaria *Brachiaria deflexa* (Schumach.) Robyns (Poaceae)

7. Horse purslane *Trianthema portulacastrum* L. (Aizoaceae)

8. Viper grass *Dinebra retroflexa* (Vahl.) Panzer. (Poaceae)

9. Field bindweed *Convolvulus arvensis* L. (Convolvulaceae)

10. Black nightshade *Solanum nigrum* L. (Solanaceae)

11. Common cocklebur: *Xanthium strumarium* L. (Asteraceae)

12. False amaranth *Digera arvensis* Forsk. (Amaranthaceae)

13. Puncture vine *Tribulus terrestris* L. (Zygophyllaceae)

14. Asthma herb *Euphorbia hirta* L. (Euphorbiaceae)

15. Carrot grass *Parthenium hysterophorus* L. (Asteraceae)

Fig. 2.14 : Common weeds in pomegranate orchard

Fig. 2.15 : Plastic mulch for water and weed management

Fig. 2.17 : Matured flowers

Male

Hermaphrodite

Intermediate

Fig. 2.16 : Male flower with rudimentary pistil; bisexual flower with long pistil; intermediate flower with medium pistil

Fig. 2.18 : A good fruit will have a high proportion of fleshy seeds in edible pulp

Chapter 3: Integrated Disease and Insect Pest Management

Fig. 3.1 : Fruit borer damaged fruit

Fig. 3.2 : Larva bores into the stem

Parasitoids

***Trichogramma* spp.**

***Tetrastichus* spp.**

***Telenomus* spp.**

***Bracon* spp.**

***Carcelia* spp.**

Encarsia inaron

Predators

Robber fly

Fire ant

Ladybird beetle

Spider

Praying mantis

Black drongo

Fig. 3.3 : Important natural enemies of pomegranate insect pests

Alfalfa | Sunflower | *Ocimum* spp.

Chrysanthemum spp. | Spearmint | Mustard

Marigold | Carrot | French bean

Cowpea | Buckwheat | Maize

Fig. 3.4 : Plants suitable for ecological engineering in pomegranate orchard

Fig. 3.5 : White flies on leaves

Fig. 3.6 Aphids on flower and shoot

Fig. 3.7 : Mealybug infestation on fruit

Fig. 3.8 : Black Scale on fruit

Fig. 3.9 : Thrips affected fruit

Fig. 3.10: Thrips affected leaves

Fig. 3.11 : Mite damaged fruit

Fig. 3.12 : Bacterial blight affected fruits

Fig. 3.13 : Symptoms of anthracnose on fruit

Fig. 3.14 : *Cercospora punicae*

Fig. 3.15 : ***Alternaria*** **fruit**

Fig. 3.16: Wilt affected tree

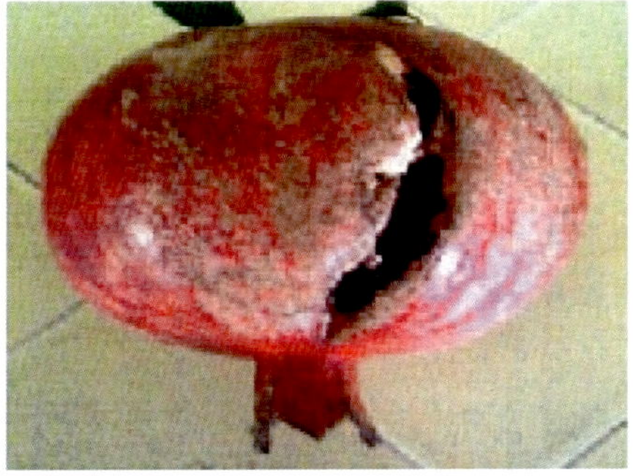

Fig. 3.17 : A view of a fruit cracking

Chapter 4: Harvesting and Post Harvest Management

Fig. 4.1 : Fruits ready for harvest

Fig. 4.2 : Packing in carton box

About the Author

S. K. Tyagi is working as a scientist (Horticulture) at Krishi Vigyan Kendra, Khargone (Madhya Pradesh). He did his B.Sc. (Ag) and M.Sc. (Ag.) in Horticulture during the year 1999 and 2001 respectively from JNKVV, Jabalpur (M.P.). He qualified National Eligibility Test (NET) twice i.e. in the year 2001 and 2006 respectively, conducted by Agricultural Scientists Recruitment Board, New Delhi. Author has more than 13 years of experience in the field of horticulture research and extension and contributed 15 research papers published in international and national peer reviewed research journals, 25 papers presented in international and national Conferences/ Symposiums/ Seminars. Nine books, 03 training manuals, 03 impact studies, 12 success stories, 10 extension bulletins and 135 popular articles go to the credit of the author. He is the editor of News Letter published by Krishi Vigyan Kenndra, Khargone. His contribution in the subject was recognised when he was bestowed with the "Best Writer Award" in 2009 by Bhoomi Nirman National News Paper, Bhopal (Madhya Pradesh). Looking to his contribution in the field he was awarded with the Young Scientist Award 2016 by IJTA, New Delhi and the Best KVK Scientist Award-2016 by Indian Society of Extension Education, New Delhi. He is the life member of various esteemed associations, advisory boards and scientific societies.

Akhilesh Tiwari is working as senior Scientist (Horticulture) at Dryland Horticulture Research & Training Center (JNKVV), Garhakota, Sagar (M.P.). He completed his M.Sc. (Ag) and Ph.D. from G.B.P.U.A&T., Pantnagar during 2002 and 2005, respectively. He started his service as Assistant Professor (Horticulture) at Haramaya University,

Ethiopia in 2005 and joined JNKVV in the same capacity at College of Agriculture, Ganjbasoda in 2008. He was selected as Senior Scientist (Horticulture) in 2012 in the Department of Horticulture, JNKVV, Jabalpur (M.P.). He has sbeen engaged in teaching, research and extension for 12 years. He has published 23 research papers in different national and international repute journals and 42 extension articles in various agricultural magazines. He has prepared 07 practical manuals, 01 bulletin and delivered various lectures of his field as resource person in conferences and worshops. Besides, he is the member of various scientific societies and committees of the Deptt. of Horticulture, Govt. of M.P. He was the recipient of Best Teacher Award by the Haramaya University, Ethiopia in 2007 and Best Paper Presentation in the International Conference at Bhopal in 2011.

Other Publications on Horticulture

S.No.	Title	Author	ISBN	Year
1	A Colour Handbook on Practical Plant Pathology	Yadav, Vijay	9789385516177	2015
2	A Colour Handbook: Landscape Gardening	Singh, Alka & B.K. Dhakuk	9789383305889	2015
3	A Handbook of Minerals,Crystals,Rocks and Ores	Alexander, P.O.	9788190723787	2009
4	A Handbook of Soil-Plant-Water-Fertilizer and Manure Analysis	Durai, M.V.	9789381450185	2014
5	A Handbook on Irrigation and Drainage	Panigrahi, Balram	9789381450888	2013
6	A Text Book of Seed Science and Technology	Padmavathi, S.	9789381450437	2012
7	Abiotic Stress and Physiological Process in Plants	A. Bhattacharya	9789385516696	2017
8	Abiotic Stress and Plant Physiology *(in 2 parts)*	A. Bhattacharya and Luxmi Vijaya	9789385516900	2018
9	Abiotic Stress Tolerance in Crop Plants: Breeding and Biotechnology	Roy, Bidhan	9788189422943	2009
10	Acid Soils: Their Chemistry and Management	Sarkar, A.K.	9789381450383	2013
11	Advances and Challenges in Agricultural Extension and Rural Development	Rathakrishnan, T.	9789380235035	2009
12	Advances in Agri-Management	Rakhi Gupta *et al.*	9789385516894	2017
13	Advances in Fruits and Vegetables Processing and Preservation	V.K.Joshi		2018
14	Advances in Nutrient Dynamics in Soil-Plant Systems for Improving Nutrient use Efficiency	Elanchezhian, R.	9789385516962	2017
15	Advances in Preservation and Processing Technologies of Fruits and Vegetables	Rajarathnam, S.	9789380235523	2011
16	Advances in Protected Cultivation	Singh, Brahma, Balraj Singh	9789383305179	2014
17	Advances in Soil Borne Plant Diseases	Naik, Manjunath	9788189422813	2008
18	Agricultural Biotechnology: Indian Print	Persley, G.J.: et al	9788189422097	2006
19	Agricultural Extension: Worldwide Innovations	Saravanan, R.	9788189422967	2008
20	Agricultural Marketing: Perspectives and Potentials	Bhat, Anil & S.P. Singh	9789385516153	2016
21	Agricultural Plant Biochemistry	Nagaraj, G.	9789383305551	2015
22	Agricultural Statistics: A Guide for Competitive Examinations	Kushwaha, K.S.	9789381450314	2012
23	Agriculture Bioinformatics	Keshavachandran, R.	9789383305421	2015
24	Agri-Food Crops: Processing,Value Addition, Packaging and Storage	R.Sasi Kumar	9789381450406	2012
25	Agri-Horticultural Biodiverstiy of Temperate and Cold Arid Regions	Zeerak, Nazir Ahmad	9789381450086	2012
26	Agrobiodiversity and Sustainable Rural Development	Soam, S.K. & M.Balakrishnan	9789383305896	2015
27	Agro-enterprises For Rural Development and Livelihood Security	Sharma J.P. & S.K.Dubey	9789380235851	2011

For details, please visit us at www.nipabooks.com

No.	Title	Author	ISBN	Year
28	Agroforestry for Increased Production and Livelihood Security	Sushil Kumar Gupta. Pankaj Panwar & Rajesh Kaushal	9789385516764	2017
29	Agroforestry Systems for Resource Conservation and Livelihood Security in Lower Himalays	Panwar,P & Dadhwal	9789381450215	2012
30	Agroforestry: Principles and Practices	Patra, Alok Kumar	9789381450765	2013
31	Agroforestry: Systems and Practices	Puri, Sunil & Pankaj Panwar	9788189422622	2007
32	Agroforestry: Systems and Prospects	C.B.Pandey & O.P.Chaturvedi	9789381450970	2014
33	Agro-Informatics	Vanitha, G & M.Kalpana	9789380235714	2011
34	Agronomy: Principles and Practices	E. Somasundaram & M. M. Amanullah	9789385516740	2017
35	Agrotechnology Manual: Including Nursery Management and Practices	Etomes, Marcel N	9789383305018	2014
36	An Introduction to Intellectual Property Rights	Pathak, Manju	9789383305124	2014
37	An Introduction to Nanotechnology	Rathinasamy, A.	9789381450413	2012
38	Ancestral Knowledge in Agri-Allied Science	Saha, Ratan Kumar	9789383305216	2014
39	Applied Computational Biology and Statistics in Biotechnology and Bioinformatics (Set of 2 Vols.)	Roy, A.K.	9789380235929	2012
40	Applied Statistical Techniques	Imam, Ekwal	9789383305537	2015
41	Applied Statistics for Agricultural Sciences	Venkatesan, D.	9789383305285	2014
42	Approaches for Incorporating Drought and Salinity Resistance in Crop Plants	Chopra,V.L. & R.S.Paroda	9789383305742	2015
43	Aromatic Plants: Vol.01. Horticulture Science Series	Skaria, B.P. et.al.	9788189422455	2007
44	Auddaniki ke Adharbhoot Sidhant Tatha Nashijeev Prabandhan	Tiwari, A.K.	9789380235998	2012
45	Bananas and Plantains: Postharvest Management, Storage,Ripening and Processing	Narayana, C.K.	9789383305452	2015
46	Basic Concepts in Statistics	Kushwaha, K.S.	9788189422400	2009
47	Basics of Horticulture: 2nd Revised and Expanded ed. (As per Revised ICAR Syllabus)	Peter, K.V.	9789383305735	2015
48	Basics of Mutation Breeding	Thirugnanakumar, S.	9789383305193	2014
49	Basics of Research Methodology	Imam, Ekwal	9789383305438	2015
50	Biochemical Aspects of Plant Physiology: Technology and Methodology	Bhattacharya, Amitav	9789383305902	2016
51	Biochemistry,Molecular Biology and Biotechnology: Instant Notes	Gajera, H.P.	9789383305520	2015
52	Bioinformatics in Agriculture: Tools and Applications	Balakrishnan, M. & Dam Roy ed.	9789381450925	2014
53	Biometrical Methods in Horticultural Science	Nirmal Sharma, V.K. Wali, Parshant Bakshi	9789385516504	2016
54	Biostatistics: Basic Concepts and Methodology	Kushwaha, K.S.	9789383305469	2014
55	Biotechnology in Horticulture: Methods and Applications	Peter, K.V.	9789381450918	2013
56	Biotechnology in India: Initiatives and Accomplishments	Niladri, Bag	9789385516252	2016
57	Biotechnology: Practical Manual Series Vol 04	Thara, K.M.	9788190851237	2009
58	Bougainvillea	S.K. Datta, R. Jayanthi & T. Janakiram	9789385516702	2017

No.	Title	Author	ISBN	Year
59	Breeding and Biotechnology of Flowers: Set of 2 Vols. (Set Price)	Singh, A.K.	9789383305612	2014
60	Breeding and Biotechnology of Flowers: Vol 01: Commercial Flowers	Singh, A.K.	9789383305353	2014
61	Breeding and Biotechnology of Flowers: Vol 02: Garden Flowers	Singh, A.K.	9789383305407	2014
62	Breeding and Protection of Vegetables	Rana, M.K.	9789380235493	2011
63	Breeding of Horticultural Crops: Principles and Practices: 2nd Revised & Expanded ed.	Kumar, N.	9789383305773	2015
64	Breeding,Biotechnology and Seed Production of Field Crops	Roy, Bidhan	9789381450680	2013
65	Cashew Production and Processing Technology: Recent Advances	Gajbhiye, R.C. & B.V. Padhiar	9789383305827	2015
66	Climate Change and Agricultural Food Production	Kibria, Golam et.al.	9789381450512	2013
67	Climate Change and Agroforestry: Adaptation, Mitigation and Livelihood Security	C.B.Pandey, M.K.Gaur & R.K.Goyal		2018
68	Climate Change and Food Security	Datta, M.& N.P.Singh	9788189422387	2008
69	Climate Change and Natural Resources Management	Lenka, & Lenka	9789381450673	2013
70	Climate Change and Plantations in the Humid Tropics	GSHLV Prasada Rao and C.S. Gopakumar	9789385516368	2016
71	Climate Change and Sustainable Agriculture	P Suresh Kumar	9789385516726	2017
72	Climate Mitigation and Carbon Finance: Global Initiatives & Challenges	Sahoo, A.K.	9789381450024	2012
73	Climate Resilient Animal Agriculture	GSLHV Prasada Rao		2018
74	Climate Resilient Crops for the Future	Peter, K.V.	9789383305599	2015
75	Climatic Variability: Impacts on Agriculture and Allied Sectors	Datta, M.:Ed.	9789381450949	2014
76	Commercial Crops Technology: Vol.08. Horticulture Science Series	Kurian, A. & K.V.Peter	9788189422523	2007
77	Commercial Horticulture	Patel, N.L. & S.L. Chawla ed.	9789385516238	2016
78	Computers in Agriculture	M. K.Sharma & Anil Bhat	9789385516160	2017
79	Crop Diseases Management: Principles and Practices	Narayanasamy, P.	9789380235677	2011
80	Crop Diseases: Identification,Treatment and Management	Darwin, Henry	9789380235462	2011
81	Developments in Physiology,Biochemistry and Molecular Biology of Plants Vol 01	Bose, Bandana & A.Hemantaranjan: ed.	9788189422028	2005
82	Developments in Physiology,Biochemistry and Molecular Biology of Plants Vol 02	Bose, Bandana & A.Hemantaranjan: ed.	9788189422929	2008
83	Dimensions of Extension Education	Mohapatra, B.P.	9789381450987	2014
84	Diseases of Field Crops and Their Integrated Management	Sanjeev, Kumar	9789385516283	2016
85	Diseases of Horticultural Crops Identification and Management: With Colour Illustrations	Kumar, Sanjeev	9789383305643	2015
86	Diseases of Vegetable Crops and Their Integrated Management:A Colour Handbook	Mishra, R.K.	9789381450499	2013
87	Drip and Sprinkler Irrigation	Biswas, R. Kumar	9789383305766	2015
88	Droughts in Agricultural Production: Monitoring & Management	Rao, G.G.S.N.	9789385516009	2015
89	Drug Discovery and Development: Traditional Medicine and Ethnopharmacology	Patwardhan, Bhushan	9788189422295	2007

90	Drying Technologies for Foods: Fundamentals and Applications	Prabhat K. Nema Arun S.Mujumdar	9789383305841	2015
91	Drying Technologies for Foods: Fundamentals and Applications: Part-II	Prabhat K. Nema	9789385516399	2016
92	Eco Agriculture Revolution	M.H.Mehta		2018
93	Ecologically Based Integrated Pest Management (Set of 2 Vols.)	Abrol,D.P.& U.Shankar	9789380235950	2012
94	Economics,Marketing and Sales of Agricultural Products	Thakur, S. Nath	9789381450659	2013
95	Efficiency Indices for Agriculture Management Research	Devasenapathy, P. & B.Gangwar	9788189422882	2008
96	Elementary Agriculture	Nainwal & Nainwal	9789385516535	2017
97	Elements of Entomology	H.Lewin, Devasahayam	9789381450635	2014
98	Emerging Technologies of the 21st Century	Roy, A.K.	9789383305339	2015
99	Encyclopedia of Agricultural Meteorology	Dhaliwal, L.K. & S.S.Hundal	9789380235547	2011
100	Enhancing Nutrient Use Efficiency	Ramesh, K	9789385516733	2017
101	Entomology: Novel Approaches	Jain,P.C. & M.C.Bhargava	9788189422325	2007
102	Essence of Horticulture	M.S. Patil, A.R. Karale,	9789385516481	2016
103	Essential Oils and their Applications	Das, Kuntal	9789381450741	2013
104	Ethnomedicinal Plants Resource of Orissa Vol 01	Sahoo, A.K.	9789380235752	2011
105	Evaluation and Impact Assessment of Technologies and Developmental Activities in Agriculture,Fisheries and Allied Fields	Roy, A.K.	9789380235400	2011
106	Experimental Biotechnology: Practical Manual Series 06	Dutta, Sunita & Abhijit Dutta	9789380235721	2011
107	Experimental Phytochemical Techniques	Raaman, N.	9789380235943	2012
108	Extension Management in the Information Age Initiatives and Impacts	Philip, H. & T.Rathakrishnan	9789381450543	2013
109	Extension of Technologies: From Labs to Farms	Anandaraja, N.	9788189422837	2008
110	Family Farming and Rural Economic Development	M.L.Choudhary & Aditya	9789383305858	2015
111	Flower Crops: Cultivation and Management	Singh, A.K.	9788189422356	2006
112	Flowering Trees: Vol.12. Horticulture Science Series	Valsalakumari, P.K.	9788189422509	2008
113	Flowers for Trade: Vol.10. Horticulture Science Series	Sheela, V.L.	9788189422516	2008
114	Food and Nutritonal Security by Sustainable Agriculture:	Mishra, Kumar Bijesh	9789383305049	2014
115	Food Engineering and Technology	Sharma, H.K. & Ashutosh Upadhyay	9789383305483	2015
116	Food Process Engineering and Technology	Pare, Aakash & Mandhyan	9789380235431	2011
117	Food Processing Waste Management: Treatment and Utilization Technology	Joshi V.K. & Satish Sharma	9789380235592	2011
118	Food Product Development and Process Innovations (in 2 parts)	H N Mishra		2018
119	Food Science	Bawa, A.S. & O.P.Chauhan	9789381450147	2013
120	Food Science and Technology: Glossary of Preeminence	Dev Raj	9789380235806	2011
121	Forest Seed Science and Management	Shukla, Gopal	9789385516757	2017
122	Forestry Science: Fundamentals and Terms	Sharad Nema	9789385516382	2016

123	Fruit Breeding	Dinesh, M.R.	9789383305513	2015
124	Fruit Crops: Vol.03. Horticulture Science Series	Radha, T. & Lila Mathew	9788189422462	2007
125	Functional Foods	Mishra, H.N. Rajesh Kapur	9789383305988	2016
126	Functional Foods and Nutraceuticals: Sources and their Developmental Techniques	Riar, C.S. & Saxena, D.C. Ed.	9789383305964	2015
127	Fundamentals of Garden Designing: A Colour Encyclopedia	Roy, Rup Kumar	9789381450307	2013
128	Fundamentals of Ornamental Horticulture and Landscape Gardening	Tiwari, A.K.	9789381450079	2012
129	Fundamentals of Vegetable Production	Rana, M.K.	9789380235707	2011
130	Genetic Diversity and Phenotypic Stability in Crop Plants	S. Thirugnanakumar	9789385516955	2018
131	Genetic Resources and Seed Enterprises: Management and Policies: In 2 Parts	Ram, Hari Har & R.Yadava	9788189422653	2007
132	Genetics and Breeding of Flower Crops *(in 2 parts)*	M.S.Patil	9789385516948	2018
133	Genomics and Genetic Engineering	Satya, Pratik	9788189422776	2007
134	Geographic Information System	Gurugnanam, B.	9788190851282	2009
135	Geoinformatics Applications in Agriculture	Singh, Anil Kumar & U.K.Chopra	9788189422233	2007
136	Geospatial Technologies for Natural Resources Mgt.	Soam,S.K. & P.D. Sreekant	9789381450802	2013
137	GIS: Fundamentals,Applications and Implementations	Elangovan, K.	9788189422165	2006
138	Glimpes of Practical's in Extension Education	K. Pradhan		2018
139	Good Management Practices for Horticultural Crops	M.K. Jatav, *et al.*	9789385516641	2016
140	Guava: Evaluation Studies of Most Promosing Cultivars	Dolkar, Disket	9789385516849	2016
141	Heterosis Breeding in Vegetable Crops	Rai, Nagendra	9788189422035	2006
142	Horticulture Science Series Vol 01-12: Set of 12 Vols	Peter, K.V.: Series Ed.	9788189422479	2008
143	ICTs for Agricultural Extension: Global Experiments, Innovations and Experiences	Saravanan, R. ed.	9789380235240	2010
144	ICTs for Transfer of Technologies: Tools and Techniques	Verma, S.R.	9788193014479	2015
145	Identification and Management of Horticultural Pests	Ranjith, A.M.	9789381450567	2013
146	Illustrated Dictionary of Entomology	Paras Nath	9788189422561	2007
147	Illustrated Dictionary of Floriculture and Landscaping	M. Kannan		2018
148	Illustrated Plant Pathology: Basic Concepts	Darwin, Henry	9789380235080	2009
149	Improving Productivity of Drylands by Sustainable Resource Utilisation and Management	Dayal, Devi	9789385516191	2016
150	Indigenous Medicinal Plants and their Practical Utility	Lakshman, H.C.	9789381450116	2012
151	Information Access in Digital Libraries	Stanley, Madan Kumar	9789383305384	2014
152	Information and Communication Technology for Agriculture and Rural Development	Saravanan, R.	9789380235882	2011
153	Information and Knowledge Management: Tools, Techniques and Practices	Roy, A.K.	9789381450628	2013
154	Innovations in Food Processing Technologies	Saiti		2018
155	Innovations in Horticultural Sciences	Peter, K.V.	9789385516344	2016
156	Innovative Horticulture	Arunkumar, K.	9788189422738	2008
157	Integrated Farming System Practices: Challenges and Opportunities	Nanda, Sankarsana	9789385516207	2016
158	Integrated Nutrient Management in Kinnow Mandarin	Bakshi, Manish	9789385516573	2016
159	Integrated Pest Management in the Tropics: In 2 Parts	Abrol, D.P.	9789385516115	2016
160	Intellectual Property Rights Demystified	Ramkumar, Mu.	9788189422875	2008

161	IPR: Drafting,Interpretation of Patent Specifications and Claims	Rathore, N.S.: Ed.	9789381450819	2013
162	Irrigation and Agricultural Drainage Engineering	Biswas, Ranajit Kumar	9789383305247	2015
163	Irrigation Systems Engineering	Panigrahi, Balram	9789380235387	2011
164	Laboratory Manual of Biochemistry: Methods and Techniques	Sengar, R.S.	9789383305025	2014
165	Laboratory Manual of Microbiology: Practical Manual Series: 05	Roy, A.K.	9789380235189	2010
166	Library Services in the Knowledge Web	Veeranjaneyulu & R.Mahapatra	9789381450192	2012
167	Management of Horticultural Crops: Vol.11 Horticulture Science Series: (Part-I & II Combined in 1 Binding)	Pradeepkumar, T.	9788189422493	2008
168	Managing Soil Health for Maximising Crop Productivity	Obeng, F.k, Avornyo,	9789385516788	2017
169	Mechanization of Cultivated Crops	Singh, Surendra	9789383305759	2015
170	Medicinal Plants: Vol.02. Horticulture Science Series	Kurian, A. & A.Sankar	9788189422424	2007
171	Merging Plant Breeding with Crop Biotechnology	Yasin, J.K.:Ed.	9789381450598	2012
172	Methods and Techniques in Plant Physiology	Bhattacharya, A. & Vijay Laxmi	9789383305506	2015
173	Microbial Biotechnology for Sustainable Agriculture, Horticulture and Forestry	Bagyaraj, D.J.	9789380235820	2011
174	Microbial Diversity and Functions	Bagyaraj, D.J. & Tilak, K. et.al.	9789381450109	2012
175	Microbial Diversity and Its Applications	Barbudde, S.B., R.Ramesh & N.P.Singh	9789381450666	2013
176	Mobile Phones for Agricultural Extension	Saravanan, R.	9789383305230	2014
177	Modern Biotechnology and Its Applications (Set of 2 Vols.)	Behera, K.K.	9789381450833	2013
178	Modern Methods in Plant Physiology	Srivastava, Girish Chand	9789380235011	2010
179	Modern Technologies for Sustainable Agriculture	Kumar, Sunil	9789381450611	2013
180	Modern Technology in Vegetable Production	Hazra, Pranab	9789380235325	2011
181	Molecular Biology and Biochemistry	Puttaraju, H.P.	9788189422257	2008
182	Molecular Markers and Plant Biotechnology	Tomar, R.S.	9789380235257	2010
183	Nanotechnology in Agriculture	Subramanian, K.S.	9789383305209	2015
184	Nanotechnology in Soil Science and Plant Nutrition	Adhikari, Tapan	9789381450789	2013
185	Nematology: Fundamentals and Applications	Jonathan, E.I.	9789380235141	2010
186	Novel Food Processing Technologies	Nanda, Vikas & Savita Sharma	9789385516047	2017
187	Numericals and Short Questions in Farm Machinery, Power and Energy in Agriculture	Yadav, Rajvir	9788190723718	2009
188	Nutraceutical Values of Horticultural Products	Dhurendra Singh	9789385516979	2018
189	Nutritional Disorders in Fruit Crops: Diagnosis and Management	Prakash, M.	9789381450956	2013
190	Objective Agriculture	Rupinder Singh	9789385516993	2018
191	Objective Entomology	Devinder Sharma	9789385516290	2018
192	Objective Horticulture	Verma, Anil	9789381450239	2012
193	Oilseeds: Properties,Products,Processing and Procedures	Nagaraj, G.	9788190723756	2009
194	Orchid: Cultivation and Management	Tiwari, A.K.	9789383305001	2014
195	Organic Farming	Singh, A.K.	9789385516139	2015
196	Organic Farming: Scope and Uses of Biofertilizers	Panwar, JDS & Amit Kumar Jain	9789385516184	2016
197	Organic Spices	Parthasarathy, V.A	9788189422844	2008
198	Ornamental Gardening and Landscaping	M. Kannan, *et al.*	9789385516689	2016
199	Ornamental Horticulture	Sindhu,S.S.ed.	9789383305865	2016

200	Ornamental Plants	Sabina, George	9788190851268	2009
201	Pest Management and Residual Analysis in Horticultural Crops	Gulati, Rachna & Beena Kumari	9789381450710	2013
202	Pesticides: Methods of Their Residues Estimation	Kumari, Beena & T.S.Kathpal	9789380235394	2010
203	Pests of Vegetables: Bionomics and Management	Laskar, Nripendra ed.	9789385516016	2015
204	Phal Evam Sabji Parirakshan Digdarshika	Kumari, Arunima	9789380235561	2011
205	Physio-Biochemistry and Biotechnology of Vegetable Crops	Rana, M.K.	9789380235318	2011
206	Physiological Disorders of Fruit Crops	Savreet, Sandhu & Bikramjit Singh Gill	9789381450581	2013
207	Phytochemical Techniques	Raaman, N.	9788189422301	2006
208	Plant Biochemistry: Techniques and Procedures	Nagaraj, G.	9789383305940	2015
209	Plant Diseases: Identification and Management (With Illustrations)	Rai, J.P.	9789383305315	2014
210	Plant Nutrient Disorders: Diagnosis and Management	Sarkar, A.K. & P. Mahapatra ed.	9789385516023	2015
211	Plant Pathogens and Principles of Plant Pathology	Kumar, Sanjeev	9789385516078	2016
212	Plant Secondary Metabolities	Shukla, Y.M. & R.Bhatnagar	9788190851220	2009
213	Plant Taxonomy and Biosystematics: Classical and Modern Methods	Rana, T.S.	9789383305414	2014
214	Plantation Crops	Bhani Ram	9789385516511	2016
215	Plants For Wellness and Vigour	V.L.Chopra	9789386546036	2018
217	Postharvest Management and Processing of Fruits and Vegetables: Instant Notes	Sharma, Satish	9789380235202	2010
218	Postharvest Techniques and Management for Dry Flowers	V.Ponnuswami & Aruna,P.	9789380235868	2011
219	Postharvest Technologies for Commercial Floriculture	Verma, Anil	9789381450048	2012
220	Postharvest Technology and Engineering: An Illustrated Guide	Dev Raj	9789381450451	2012
221	Postharvest Technology and Processing of Horticultural Crops	Pandit, P.S.	9789383305278	2014
222	Postharvest Technology of Horticultural Crops: Practical Manual Series Vol 02	Sharma, Satish & M.C.Nautiyal	9788190851206	2009
223	Postharvest Technology of Horticultural Crops: Vol.07. Horticulture Science Series	Sudheer, K.P. & V.Indira	9788189422431	2007
224	Practical Manual of Entomology (Insects and Non-Insects Pests)	Devasahayam,H.Lewin	9789380235905	2011
225	Practical Manual of Horticulture Crops: Set of 2 Vols.	Verma, Anil Kumar	9789383305711	2015
226	Practical Manual of Horticulture Crops: Vol 01: Production Technologies	Verma, Anil Kumar	9789383305704	2015
227	Practical Manual of Horticulture Crops: Vol 02: Processing and Postharvest Technologies	Vaidya, Devina	9789383305698	2015
228	Practical Plant Biotechnology and Genetics	Rani, Archana	9789383305995	2015
229	Precision Farming in Horticulture	Singh, Jitendar, S.K.Jain,L.K.Dashora	9789381450475	2013
230	Production Technology of Spices,Aromatic,Medicinal and Plantation Crops	Barche, Swati	9789385516061	2016
231	Propagation of Horticultural Crops: Vol 06 Horticulture Science Series	Rajan, S. & B.L.Markose	9788189422486	2007
232	Propagation of Horticultural Plants: Arid and Semi-Arid Regions	Singh, R.S. & R.Bhargava	9789383305254	2014

	Title	Author	ISBN	Year
233	Protected Cultivation of Horticultural Crops	Singh, D.K & K.V.Peter	9789383305155	2014
234	Quality Control for Value Addition in Food Processing	Dev Raj, *et al.*	9789380235578	2011
235	Quantitative Genetics and Crop Breeding	Thirugnanakumar, S.	9789380235981	2012
236	Question Bank in Extension Education	Mohammad, Asif & K. Ponnusamy	9788193014493	2015
237	Question Bank in Forest Science	Gopal Shukla		2017
238	Question Bank in Fruit Science	Jarande, S. D	9789385516795	2017
239	Question Bank in Postharvest Technology	Guleria, SPS & Anil Kumar Verma	9789380235042	2010
240	Recent Advances in Biopesticides	Johri, Jayendra: eds.	9789380235219	2010
241	Recent Trends in Horticultural Biotechnology: In 2Vols	Keshavachandran, R.	9788189422592	2007
242	Rejuvenation in Fruit Trees: Tropical and Subtropical	Kumar, Rajesh	9789385516436	2016
243	Remember Your Humanity: Pathway to Sustainable Food Security	Swaminathan, M.S.	9789381450178	2012
244	Rural Livelihood and Food Security	Wani, M.H.	9789380235936	2012
245	Seed Processing: A Practical Approach	Rakesh C. Mathad	9789385516085	2018
246	Seed Production of Field Crops	Mondal, S.S.	9788190723763	2009
247	Seed Science and Technology	Vanangamudi, K.	9789383305117	2014
248	Seed Testing Techniques for Seed Spices	Sangeeta Yadav & Arun Kumar barholia	9789385516917	2018
249	Soil Conservation: Fully Revised and Updated: 3rd ed.	Hudson, Norman	9789383305971	2015
250	Soil Microbiology and Biochemistry	Hassan, G.Dar	9789380235134	2010
251	Soil Sampling and Methods of Analysis	Pal, Sushant	9789381450574	2013
252	Soil Science: An Elementary Textbook	Puri, A.N.	9789383305063	2015
253	Soil Testing and Analysis: Plant, Water and Pesticide Residues	Brajendra, Patiram	9788189422707	2007
254	Spices,Plantation Crops,Medicinal and Aromatic Plants: A Handbook	Tyagi, S.K.	9788193014486	2015
255	Spices: Vol.05. Horticulture Science Series	Nybe, E.V. and Mini Raj	9788189422448	2007
256	Statistical Designs and Analysis for Agricultural Field Experiments	Katyal, Vijay & D.M.Hegde	9789381450840	2013
257	Statistical Methods for Agricultural Field Experiments	Katyal, Vijay & B.Gangwar	9789380235424	2011
258	Sustainable Agriculture: A Vision for Future	Desai, B.K. and B.T.Pujari	9788189422639	2007
259	Sustainable Environmental Science	Sahu, D.D.	9789381450208	2012
260	Systematics of Fruit Crops	Sharma, Girish	9789380235066	2009
261	Tea: Technological Initiatives	Niladri Bag, Arundhati Bag and L.M.S. Palni	9789385516337	2016
262	Technologies for Sustainable Green Environment	Davamani, V.	9789381450420	2012
263	Temperate Horticulture: Current Scenario	Kishore, D.K.: et. al.	9788189422363	2006
264	Textbook of Floriculture and Landscaping	Singh, A.K.	9789386546005	2017
265	The Basics of Human Civilization: Food, Agriculture and Humanity Vol.01 Present Scenario	Prem Nath	9789381450734	2013
266	The Basics of Human Civilization: Food, Agriculture and Humanity Vol.02 Food	Prem Nath	9789383305377	2014
267	The Chemistry of Soil Processes	Greenland, D.J.	9789383305926	2015
268	The Pomegranate	Hiwale, S.S.	9789380235158	2009
269	The Qur-anic Plants and Animals	Zeerak, Nazir Ahmad	9789383305780	2015

270	The Science of Horticulture Vol 01	Peter, K.V.: eds.	9789380235479	2011
271	The Science of Horticulture Vol 02	Peter, K.V.: eds.	9789380235486	2011
272	The Theory of Sample Surveys and Statistical Decisions	Kushwaha, K.S. & Rajesh Kumar	9788189422899	2009
273	The Weeds of Kumaun Himalayan Region (Uttarakhand)	T.S. Rana & Bhaskar Datt	9789385516474	2016
274	Traditional Agricultural Practices: Applications and Technical Implementations	Rathakrishnan, T.	9789380235028	2009
275	Tuber and Root Crops: Vol.09. Horticulture Science Series	Palaniswami, M.S. & K.V.Peter	9788189422530	2008
276	Turning Plants Into Medicines: Novel Approaches	T, Parimelazhagan	9789381450468	2013
277	Underutilized and Underexploited Horticultural Crops: Vol 01	Peter, K.V.: ed.	9788189422608	2007
278	Underutilized and Underexploited Horticultural Crops: Vol 02	Peter, K.V.: ed.	9788189422691	2007
279	Underutilized and Underexploited Horticultural Crops: Vol 03	Peter, K.V.: ed.	9788189422851	2008
280	Underutilized and Underexploited Horticultural Crops: Vol 04	Peter, K.V.: ed.	9788189422905	2008
281	Underutilized and Underexploited Horticultural Crops: Vol 05	Peter, K.V.: ed.	9789380235288	2010
282	Value Addition in Flowers and Orchids	De, L.C.	9789381450123	2011
283	Vegetable Crops: Genetics Resources and Improvements	Singh,D.K. & H.Choudhary	9789380235509	2012
284	Vegetable Crops: Vol 04. Horticulture Science Series	Gopalakrishnan, T.R.	9788189422417	2007
285	Vegetable Seed Processing	Mathad, Rakesh C. & Basave Gowda	9789385516030	2015
286	Water Management for Enhancing Water Productivity	N.K.Gontia & H D Rank	9789385505728	2018
287	Weed Science	Das, P.C.	9789383305261	2015